Glossary

IWA	Inland Waterways Association
BWB	British Waterways Board
TWA	Thames Water Authority
LVRPA	Lee Valley Regional Park Authority
AWA	Anglian Water Authority
SWA	Southern Water Authority
Pound	Stretch of water between locks
Winding Point	Part of canal which has been widened to enable a seventy foot boat to turn round

Canals are my life

Iris Bryce

Kenneth Mason

British Library Cataloguing in Publication Data
Bryce, Iris
Canals are my life.
1. Canals — England
I. Title
386'.46'0924 HE 436

ISBN 0-85937-277-4

Published by Kenneth Mason, Homewell, Havant, Hampshire

Produced by Articulate Studios, Emsworth, Hampshire

Printed by Redwood Burn Limited, Wiltshire

This book is dedicated to a
much loved friend,
Jimmy Asman —
Grateful thanks
for many years of
encouragement and faith

Canals are my life

Contents

Illustrations

between pages 32 and 33 and 64 and 65

Introduction

Looking back on the last few years, I often wonder how on earth I managed to come to the decision that was going to change my life so dramatically. There I was settled with my husband Owen in the Kentish countryside where we enjoyed life as small farmers, with a lovely farmhouse home, plenty of good friends and our three married daughters and son living near enough to visit us frequently.

During the warm summer days as we sat around the swimming pool with family or friends and helped ourselves to strawberries picked from the garden, I could not help thinking that life was indeed good to us.

We ought to have been content for, apart from farming, Owen and I had other interests. I wrote articles and he lectured. Additionally, as a jazz musician, he taught and played regularly all over the country. So we had much to do and to think about.

And then came the day when our insurance agent advised us to have our property revalued for insurance purposes — it was still insured for its original cost of £5,000. When we received the new valuation we thought there had been a typist's error, but a telephone call to the estate agents confirmed that the incredible figure of £45,000 was correct. It did not take us long to realise that here was an opportunity to sell a house that was now far too big for the two of us and, although we enjoyed farming, it did entail a seven-day-a-week job with little time to relax. But, if we sold up, where would we go? What would we do with ourselves?

Out of the blue, Owen suggested, 'why don't we build a narrow boat and live on the canals?' We were both keen boaters but had spent only a week at a time on a boat for holidays . . .

So the die was cast and a narrow boat was built to our specification, with room on board for a piano! The boat-builder accepted the challenge. So a music room was built and the upright piano — a family heirloom — had to be craned on board before the roof could even be put on.

We had created such a stir by this seemingly crazy request of ours that the press and television cameras covered the instalment of the piano. Appropriately, we had christened our boat *Bix* after the famous jazz trumpeter Bix Beiderbecke.

A new chapter in our life began. Once the shelter of the boatyard and its company of experts lay behind us as we nosed our way on to canal waters, we realised how very much on our own we were.

It was not long before we had our first taste of the toughness of our new life when I had to use all of my seven and a half stones against the balance beam at our first lock gate. Then came another and another as I sweated and pushed. And this was just the beginning . . .

The solitude of our new world was dramatically brought home to us on our first Christmas on board. We woke on Christmas morning to find that the electricity had failed! And as we sat in the candlelight, unwashed and eating our breakfast in silence, I realised that this was the first time in 29 years that there were only two of us . . . no family, no friends.

We were determined, however, to succeed in our new life, in any case it was only for two years, for by that time we would have completed our voyage of all of England's waterways and be ready to find that cottage with roses round the door and settle down again!

Well, here we are seven years later, still on the move on our beloved *Bix*. We have had so many wonderful adventures, seen so much beauty of the English countryside, met so many fascinating people, that we are now well and truly converted to our floating nomadic life.

When I wrote my first book *Canals are my home,* I imagined that it would be a record of a two-year experiment in living. Here I am having just completed another book which surely goes to show that canals are no longer just a home but now a way of life.

1

It was a mild November morning on the canal. There was not another human in sight to disturb our tranquility, just us and a flock of redwings searching for berries on the hawthorns. Slowly and quietly we drifted through the countryside on the outskirts of Macclesfield as I sat and pondered where we should spend the winter.

We had lived on board our narrow boat *Bix* for two years, developing a routine of finding temporary work in the winter months and travelling and lecturing during the rest of the year. We had spent the previous winter on the Gloucester and Sharpness Canal and now in November we had to decide this year's base.

My day dreaming was shattered by a hail from the side of the canal.

'Hey, *Bix,* I've a message for you.'

We were passing a line of moored boats with a small boatyard at the end; a notice-board revealed that the place was called Higher Poynton. I slammed the engine into reverse while my husband Owen, also hearing the shout, hastened from below.

My heart worked overtime for many people on the canals knew how to contact us in emergency and my thoughts immediately raced between my mother, now in her eighties, the children and the grandchildren.

As I nosed the boat towards the canal-side, a young man emerged from a hut.

'Ron Young wants you to ring him,' he called. 'Saw him yesterday, he's been looking for you.'

Ron Young? Who's he, we wondered as we edged past the yard.

'What's his number?' Owen called.

We were now almost under the bridge.

'He's at Anderton Marina, on the Trent and Mersey,' he managed to shout before disappearing from view.

Anderton Marina did the trick. Owen recalled meeting a Mr Young who managed a hire-fleet of some size so when we reached Bollington later in the day he phoned the Marina . . . which is how we came to spend that winter on the Trent and Mersey Canal with the Youngs at Anderton.

We were both offered work that would last until April: Owen on full-time painting and varnishing the 25 hire boats; I would help Mrs Young on a part-time basis with the cleaning, storing and checking of the linen, cutlery, crockery, cookers and refrigerators.

But first we had to get there. By road it was a matter of an hour or so; by boat it was to take us a week! As we negotiated the locks between Kidsgrove and Wheelock, known to the old boatmen as Heartbreak Hill, I cursed once more the Rochdale Canal Company. Just ten days previously we had been only 27½ miles from Anderton Marina — having decided to cruise the Bridgewater and Trent and Mersey Canals to see the famous Anderton lift which carries boats from the Trent and Mersey 50ft down to the River Weaver. What we had not reckoned on, however, was that the Rochdale Canal — the link between us and the Bridgewater — would be closed!

The Rochdale Canal is privately owned so does not come under the administration of either the British Waterways Board or the Bridgewater Canal. When the Rochdale Company decided to close the nine locks that join the two canals just five minutes before we reached them, we were forced to turn back and travel 63¼ miles of a circular route to reach Anderton. I found it infuriating to have to work through 82 locks instead of ten.

When a survey was carried out asking holiday-makers what was the most enjoyable part of their canal holiday, working the locks came top of the list. However, in the winter months, with a crew of one it is an onerous job — although we cannot complain as canalling is the life we have chosen . . . but it would be nice to know that one could rely on co-operation now and again from the powers-that-be.

Originally, when we embarked on what was to be a two-year cruise of the canals of England, it was envisaged as a fairly light-hearted affair. 'A two year giggle,' as one young friend described it. We had planned a leisurely look at not only the waterways but also the surrounding villages and cities, the contrasting scene between town and country, and, of course, meeting so many people from so many different regions. I certainly did not expect to find myself pulling a trolley containing 50 pillow-cases for a couple of miles to a launderette and then back again, to stand and iron them — and then repeat the process the following day with another load, the next day and the next.

The only variation to this routine was that the trolley sometimes contained sheets, tea towels or dusters instead of pillow cases. In an effort to relieve the boredom of my chores, I would vary the route through the village of Anderton, along the main road, and then into Barnton, where the launderette was situated. Better still I could walk along the towpath and through the woods, although pushing a trolley this way was difficult as there was so much mud about and the path through the woods was steep. However, it did take me past the Anderton Lift, an incredible piece of engineering, built by Leader Williams in 1875, the man also responsible for the Manchester Ship Canal. The Lift carries boats up and down from the canal to the River Weaver 50ft below and as I gazed I hoped not only to find out more about it, but also to see it working or, better still, take *Bix* down to the Weaver on it.

As the weeks went by I completed the laundry work and was promoted to the washing, drying, counting and storing of endless kitchen ware for the Young's hire fleet. Nevertheless I still found time for daily walks along the canal. Within easy reach was my favourite spot: Marbury Park, a one-time private estate now open to the public as a country park. Without Marbury my winter at Anderton would have been unbearably protracted . . . I don't think I could have survived.

Neither Owen nor I fitted into the life-style of Anderton, Barnton or the town of Northwich, which was the nearest large shopping centre. We were in a tight-knit community and the very first day we boarded a bus to shop in Northwich all conversation stopped. If we had been little green men from outer space we surely could not have received a more puzzled, questioning reception.

After a week or so I began to receive a few 'good mornings' on my way to the launderette, but throughout our stay we were never really accepted. It was to be three weeks later, on my regular weekly shopping trip, before a woman actually spoke to me as I was about to leave the bus.

'You live at the marina don't you?' she asked.

I agreed we did.

'Settled in all right?' she enquired.

We carried on a conversation happily as we walked to the market and shops. However, when she realised we were staying only until April and that we had left our family behind in Kent, she backed away as if I had small-pox!

She simply could not understand how we could leave the children; the story had clearly spread when next day I visited the little Post Office stores in Anderton. Here I was stopped and asked, 'How *could* you leave your family?' Another comment made to me was, 'I could not live if I didn't have the grandchildren around every day,' and *that* was from a man! All in all we seemed to be in a world far removed from our life in the south of England.

However that winter had some compensations: every weekend we took *Bix* out of the marina, sometimes to travel along the canal to Lostock Graham and Broken Cross, or in the opposite direction towards Barnton and Dutton.

The first route was through somewhat bleak landscape interspersed with a few small woods and farms. Much of the area is mined for salt and at Marston we passed a group of derelict buildings which had once been a salt works; one building near the canal had sunk so low into the ground that the architrave above the door protruded only two feet above the surface. Another building had lost its outside wall and, as I poked my head inside, I discovered pieces of salt formed into fantastic outlines. I bore four or five of the most attractive shapes back to the boat as I knew our family would treasure them as unusual ornaments which would not have disgraced a modern art gallery.

A year later I heard a talk on the radio and learned that the 'derelict' salt works we had visited were still operating and enjoying good business supplying the growing number of health shops. The speaker explained how salt production still used methods originated by the first owners of the mines in the eighteenth century. What he did not add, however, was that most of the buildings must have belonged to the same period as well.

The scenery on our alternate weekend cruise was much prettier, especially after we had negotiated the Barnton Tunnel where the River Weaver runs parallel in the valley below.

Before we left Anderton in April we had some of our family to stay for Easter whereupon we scored a few good marks from the locals! I was not sorry to say goodbye to that part of the canal, although I knew I would miss my afternoons' bird-watching in the hides at Marbury Park. A week before our departure, I was lucky enough to help the warden count 17 pairs of Great Crested Grebes, a species of water fowl that is usually so shy that it either dives below the surface the moment a human appears, or hides in the reeds. But from January onwards, when they are seeking mates, their long white necks and chestnut feathers can be seen quite easily as they stretch upwards to show off, or fluff out their feathers and swim low and squat along the water to impress their chosen partners. Their courtship display is beautiful, if you don't mind putting up with the croaks that accompany it. I thoroughly recommend a stop-over at this country park for all boaters, whatever the season. It is easy to moor alongside, and the walks through the woods are clearly signposted. The mere itself is lined on one side with a thick fringe of reeds, the perfect habitat for the nesting and breeding of many varieties of water fowl.

We had planned that on leaving the area our first trip would be down the Anderton Lift. Throughout the winter I had waited patiently for a boat to descend but it was not until early April that I saw a British Waterways Board working boat go down. Obviously this was because no regular cargoes are carried now by water and it is the holiday-makers who provide the lift's main source of income.

It was on a cold, windy Easter Saturday that we made our way slowly into the huge water tank that was closed behind us by a guillotine-like partition. There are two tanks, each holding 250 gallons of water. Originally the lift worked hydraulically. The tanks were counter-balanced and, in order to raise a boat, water was added to the top caisson, to give the tank its technical name. So successful was the Anderton Lift that it was copied by engineers on the Continent and in Canada. Today the lift is operated electrically and each tank is independent.

It was a thrill to go down, down, down, until the huge ICI soda ash works that line the banks of the Weaver for miles, cast its shadow as well as its clouds of unpleasant but non-toxic fumes over us.

We spent a week on the river Weaver, first cruising the eleven miles to Winsford Bottom Flash, the name given to the large expanse of water caused by the land subsiding above the salt mines. The Flash is huge and with a background of woods it makes an attractive leisure centre for the local inhabitants. It is also the end of navigation and as we turned to retrace our journey

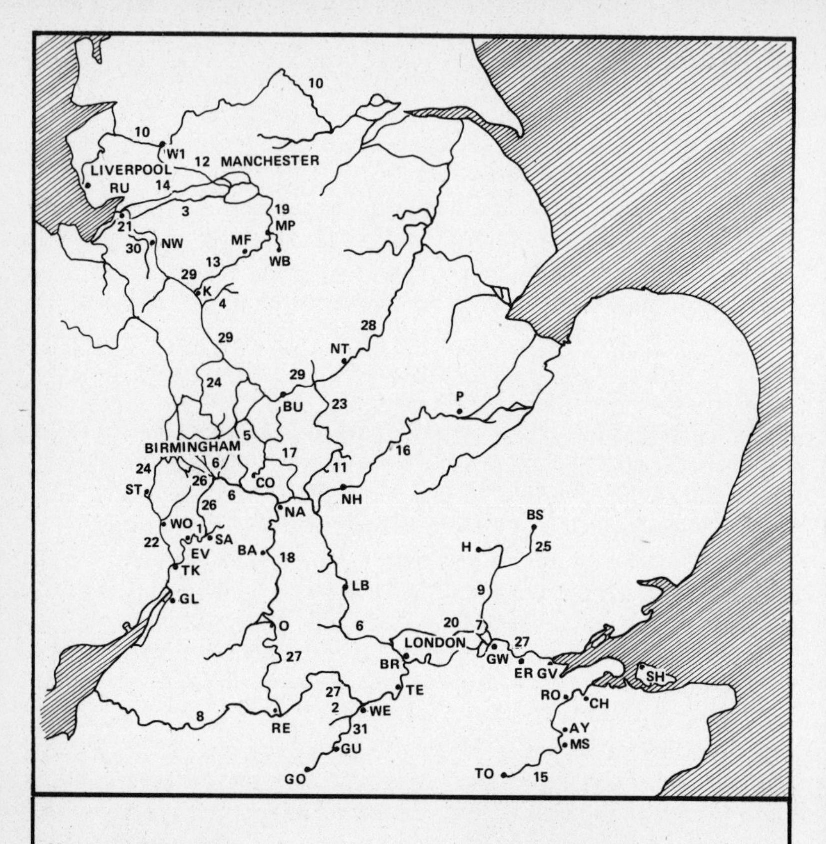

Key to waterways

1 Avon, River
2 Basingstoke Canal
3 Bridgewater Canal
4 Caldon Canal
5 Coventry Canal
6 Grand Union Canal
7 Hertford Union Canal
8 Kennet and Avon
9 Lee River
10 Leeds and Liverpool Canal
11 Leicester Arm
12 Leigh Branch
13 Macclesfield Canal
14 Manchester Ship Canal
15 Medway, River
16 Nene, River
17 North Oxford Canal
18 Oxford Canal
19 Peak Forest Canal
20 Regents Canal
21 Runcorn Arm
22 Severn, River
23 Soar, River
24 Staffordshire and Worcestershire Canal
25 Stort, River
26 Stratford Canal
27 Thames, River
28 Trent, River
29 Trent and Mersey Canal
30 Weaver, River
31 Wey, River

we realised that although many sailing boats were moored on the water we were the only vessel moving on this bitterly cold and windy day. Unfortunately there were no signs to warn of shallows and as we turned we hit a sand bank. We were firmly stuck amidships with not another soul in sight. We tried pushing off the boat with poles but without success. Tired and cold we went below for a hot drink when, through the window, I saw my first cormorant atop a post beneath the trees lining the banks.

After a further unsuccessful attempt to get clear with engine and poles Owen suggested I turn on all taps to empty our water tank, thus lightening the nose of the boat and enabling us to swing her round and off the sand. I was horrified. Water taps to replenish our supply were not that frequent on the Weaver, but of course I obeyed my captain. He was right. With the tank emptied and spare gas cylinder and other heavy weights moved to the stern, we eventually slid off and retreated hastily back down river.

At Vale Royal Locks the lock-keeper obligingly filled our portable two-gallon water container with which we had to make do until we reached Hunts Lock where we found a handy tap from which to refill the tank.

Northwich provides convenient mooring with the town centre only five minutes walk from the wharf. Their new shopping precinct is so well designed that it does not mar the original layout of the town. There I stocked up and on returning to the boat found we had acquired neighbours. Another narrow boat, painted deep maroon, an unusual colour for a canal boat, and

Key to towns

AY Aylesford
BA Banbury
BI Birmingham
BS Bishops Stortford
BR Brentford
BU Burton on Trent
CH Chatham
CO Coventry
ER Erith
EV Evesham
GL Gloucester
GO Godalming
GV Gravesend
GW Greenwich
GU Guildford
H Hertford
K Kidsgrove
LB Leighton Buzzard
LP Liverpool
LD London
MS Maidstone
MF Macclesfield
MC Manchester
MP Marple
NA Napton
NH Northampton
NW Northwich
NT Nottingham
O Oxford
P Peterborough
RE Reading
RO Rochester
RU Runcorn
SH Sheerness
ST Stourport
SA Stratford-on-Avon
TE Teddington
TK Tewkesbury
TO Tonbridge
WE Weybridge
WI Wigan
WB Whaley Bridge
WO Worcester

named *Amelia Di Liverpool,* had tied up behind us.

Her owners soon arrived laden with shopping and, in the true tradition of all boaters, it was not long before they were having coffee with us on *Bix* and we in turn were being shown over *Amelia.* Their boat was used solely as a mean of relaxation, a place where they could get-away-from-it-all and as the owner turned out to be the famous singer, John Shirley-Quirk, we could understand why he looked forward so much to the few days he could spend in this peaceful other world of canals. With his wife and family John was on the Weaver for two days before having the boat moved to a mooring in the south where he hoped to be able to use it more often. The next morning we filmed *Amelia Di Liverpool* as she entered the Anderton Lift and rose majestically to the Trent and Mersey Canal.

We continued down river. Although the waterway is referred to as a river, the maps show it as the Weaver Navigation. In some parts it is a channel built alongside the actual River Weaver, which itself is navigable for part only of the way.

Soon we were passing the vast ICI complex but, as the river swept round a bend into the Barnton Cut, the industrial scene ended abruptly and we found ourselves cruising in a valley of all shades of green.

Through Salterford Locks we were accompanied by a commercial craft owned by ICI — its orange and black livery standing out boldly against the many large ships that had come up river from the Manchester Ship Canal. We moored for the night near the swing bridge at Acton and after such a cold, windy journey were glad of an early night.

The next morning a gusty wind strengthened as we headed into it towards Frodsham where the river branches to the left. The main navigation is now marked on the maps as the Weston Canal — a canal however that is far removed from the rest of the system, as it still carries ships of 600 tons. We saw none of these monsters that day to my relief as the strong wind was playing havoc with our steering. At times Owen found it impossible to stand upright and hold the tiller. I had done some washing that morning and found myself standing at the front of the boat, clutching at wildly fluttering underwear as it tried to leap from the line and disappear up river. As I repegged it the wind tore savagely at my hair and I wondered once again why I had opted for such a life. Why, indeed, had I given up a comfortable home to career around wretchedly cold and windy rivers, not to mention weed-choked, badly maintained canals? Nevertheless the wind was good for drying, but oh, how bitterly cold it was.

Woods and fields gave way to the ugly outlines of the chemical works whose feeder pipes accompanied us to Weston Point, which we reached at one o'clock. Here the Weaver Navigation ends and locks descend to the Manchester Ship Canal.

However, we intended retracing our steps to the Anderton Lift and so to the Trent and Mersey. A dredger was working at the end of the river. As we prepared to turn, one of the men called, 'thought I didn't recognise your flags,' and he pointed to my multi-coloured underwear on the line. He was very helpful however as we struggled to reverse our course in the high wind. Eventually we secured to his hawser which stretched alongside the towpath, so fatigued from our battle with the elements that we asked if we could stay for the rest of the day and the night. Even the fumes from our neighbours' smoking chimneys, a few feet away, seemed not to matter to our weary minds and bodies.

'Yes, of course you can,' was the answer. 'There won't be anything moving until tomorrow. I was right surprised to see you. They gave out over the radio that all shipping was to stop due to gale force winds. Don't suppose you carry radio though, do you?'

They do say ignorance is bliss.

Next day I found an answer to my question, 'why, oh why, am I living like this?' The sun shone from an azure sky on dark grey water which reflected back a rippling silver, even the rows of white pipes adorning the chemical works shimmered appealingly.

Our return journey was magical. Re-entering the Weaver Valley the blue sky and green fields, framed within the stone arches of the railway viaduct straddling the river at Dutton, looked quite unreal. We felt part of a painting — a scene which I had certainly missed in our frantic journey downstream.

Ascent by the Anderton Lift is far more dramatic than descent; in the stern of our boat we felt as if we were preparing for take-off as we slowly rose while below the ships, loading at the ICI wharf, grew ever smaller.

Soon we were back again on our normal waters, this time heading for where it all began in the heyday of the canal age — the Bridgewater Canal at Worsley.

2

Where does one canal begin and another end? If it is at a junction such as the North Oxford with the Coventry, or the Coventry with the Trent and Mersey then it is easy. Some canals are separated by a stop-lock; these usually have a fall of a few inches only, a well known device for older canal companies to steal precious water from newcomers. But where do the Trent and Mersey and the Bridgewater Canals start and finish?

As you approach the end of the Trent and Mersey there are three tunnels to navigate between Barnton and Dutton. The first, Barnton, is 572 yards long and its slanted entrance onto the canal makes steering into it difficult. Salterford, the second tunnel, is only a few hundred yards further along the canal and this, like Barnton, has a few interior kinks and curves, but once through its 424 yards beautiful scenery graces the next five miles to Preston Brook Tunnel which at 1239 yards is the longest of the three. When you merge into the open a notice by the side of the canal informs you that you are now on the Bridgewater Canal. So, apparently, the systems meet at the end of the tunnel.

Canals, like roads, carry signposts but each canal company has a different design. Trent and Mersey signs are divided in half, each section stating the start and finish of the canal and the appropriate distance you are from either end. So if for instance you are at Fradley the signpost will read 'Shardlow 26 miles' on one half of it and 'Preston Brook 67 miles' on the other . . . Shardlow and Preston Brook being at opposite ends of the canal.

We moored for lunch just outside Preston Brook tunnel and I seized the opportunity for a walk. It was during this stroll that I discovered the only signpost I have ever seen that actually marks the beginning of a canal. It read 'Shardlow 93 miles' on one half and 'Preston Brook' on the other; so here beginneth the Trent and Mersey.

I have a photo of the sign which took a lot of finding; you certainly won't see it if you are on a boat although it is in the perfect place to be spotted by the horse which used to haul the boat. However it is typical of the sort of incident that makes it so worthwhile leaving the boat now and again and taking a walk along the towpath for a change.

The Bridgewater Canal is owned by the Manchester Ship Canal company and visiting boats licenced by the BWB may use it for a week at a time. The oldest example of our canal system, it was the brainchild of the Duke of Bridgewater who, in the eighteenth century, developed the idea of using boats to carry the coal from his mines at Worsley to Manchester.

James Brindley was the engineer, and from its opening in 1761 the canal was a tremendous success. The work involved in its construction must have meant at least chaotic if not appalling conditions for the local villagers, farmers and tenants. We appreciate what happens when a motorway is built near our home, but imagine the army of labourers who would descend on the community for months at a time while they dug and blasted their way through the countryside. With pick, shovel and wheelbarrow their work was long and arduous. They were called 'navigators' because they were digging a channel for navigation and so the word 'navvy' passed into our language.

The Bridgewater Canal saw the construction of the first swing aqueduct at Barton, an engineering feat comprising a trough of water that can be swung to one side to allow ships to pass below on the Manchester Ship Canal.

Where the Bridgewater starts an arm leads to Runcorn, a journey of six miles or so with a spectacular view of the Widnes Road Bridge which spans three waterways — the Runcorn Canal, the Manchester Ship Canal and the River Mersey. Runcorn was designated a New Town in 1964 and of the seventeen bridges (not including railway bridges) that cross this short canal no fewer than four are called Runcorn New Town Bridge!

Back again on the main canal, we were delighted to find an industrial site whose owners had taken the trouble of landscaping their grounds to blend with the countryside and the canal ambience. We stopped to photograph the trim lawns and the shrubs that almost concealed railings, transformers and bleak walls. The canal bridge road led to the site where an adjacent sign informed us that this was the Daresbury Nuclear Physics Laboratory. We crossed the bridge to the village of Daresbury where we were greeted with another pleasant surprise: a church whose stained glass windows contain figures of the characters from *Alice in Wonderland.* They include the Mad Hatter, the White Rabbit, and even Alice herself. As Lewis Carrol was a former incumbent of Daresbury church these windows form a fine memorial to a man who brought so much magic into the lives of children and adults alike. I can well imagine during prayers how many a tousle-haired fidgety toddler has peeped through his fingers to see a rosy Alice smiling down at him.

The 20 mile stretch from Daresbury to Stretford passes through pleasant countryside. We stopped at Lymm, where moorings were excellent and the town itself was attractive with its well-kept market cross, dating back several centuries. The canal crosses three aqueducts between Preston Brook and Stretford, that over the River Bollin having been repaired after a disastrous breach in 1971.

Although there were many boats moored on this stretch of water, (Lymm Cruising Club alone had three separate mooring sites and the Sale club two) we were to meet with no other boaters during our four day trip.

At Stretford the canal divides at a junction called Waters Meeting where the main canal turns off to Leigh. The other section continues into Manchester to join the Rochdale Canal at the bottom of the flight of nine locks which we recalled had been closed to us when five months earlier we had reached them from the other side. We had no intention of paying the Rochdale Canal Company to use the locks now so that we could complete our itinerary and be able to claim one day that we had voyaged over every part of the canal system navigable to us. We selected the Leigh Branch instead and what a scenic change lay ahead.

A vast industrial estate adjoining Barton Docks gave way to the Barton Power Station after which we found ourselves on the Barton Swing Aqueduct from which we had a sensational view below of the enormous wide Manchester Ship Canal. A large vessel was moving steadily towards the aqueduct, coaxed by two fussy tugs alongside, so once we were on the other side, near the Worsley Cruising Club, we moored. Out came our cameras as both the aqueduct and the road bridge swung round to allow the ship to pass beneath. We were too near the aqueduct to photograph the entire scene but Owen climbed on to a wall and managed to obtain a reasonable shot of this giant tank, carrying 800 tons of water, suspended in mid-air.

A mile further on we reached Eccles where of course I just had to buy some real Eccles cakes. At one time I worked in a bakery and sold hundreds of these popular pastries, but the home-town product was a real suprise. The cakes were smaller, only about two and a half inches in diameter instead of the usual four inches; made with butter they were also spicier. I spent quite a time looking for a shop that sold them and after a walk around the back streets discovered them in a small grocer-baker-general store, a worthwhile find, and not simply because of the cakes.

As I left, I noticed on the outside of the shop windows the words 'Fresh Ground Coffee' at a ridiculously low price. Coffee beans were a luxury to us now but surely if they advertised 'fresh ground' they must stock beans. Yes, they did and I bought half a pound. If they were fresh and not something recently discovered at the back of a shelf, then I would buy more on our return.

Breakfast next morning yielded a nostalgic delight; the smell of the beans as we ground them, and the aroma from the coffee pot brought memories of other days, another life! It *was* good coffee so I went back for three pounds more. Alas, the assistant had paper bags which held but half a pound. As I watched the number of bags increase and the beans in the sack diminish, I asked how much was left in stock. She looked at me strangely, quite convinced by now that I was mad! I am sure she had not sold beans alone before, and I suspect that the sack had been ordered by mistake in the first place, as there was no sign of a coffee-grinding machine on the premises.

She had seven pounds remaining so, as the price was well below average, I decided to take the lot.

'Put them back in the sack and I'll carry them in that,' I said.

The poor woman glanced furtively at me as she emptied each paper bag into the sack. And then I paid by cheque! At least I *tried* to pay by cheque, but she shook her head and disappeared into the back of the shop. Out came the owner, a surprisingly young man, rather fat and wearing a splendid brilliant white old-fashioned grocer's apron, the sort that comes down well over the legs and has a bib front.

He too refused my cheque, despite my banker's card.

'We did take one once but no more,' he said, shaking his head and directing me to a bank about a quarter of a mile into town. It was worth the walk however as I cheerfully anticipated the savour of our breakfast during the following few weeks.

3

On Worsley Bridge we looked down on water that could well have been poured from a tomato soup can; it was bright orange! So this was where the young Duke of Bridgewater had started what was to become known as the Canal Age. Thus said a notice on the bridge visible to motorists and pedestrians, but not to boaters. If you happen to think canalling is a test of seeing how many miles you can travel in a set time, you could easily zoom through Worsley without appreciating its importance in canal history.

It is still possible to see the entrance to the tunnels built by the Duke for not only did he use boats to deliver coal but he also had a fleet of specially-designed smaller boats, called 'starvationers', that travelled through 40 miles of underground canals, hauling coal from mine to surface.

A picturesque black and white house with steps leading to the water stands by the canal basin. Every tourist frantically photographs this thinking that it is the Duke of Bridgewater's house whereas it is in fact The Court House; the Duke's former home is a mile away.

We walked over to see this partly-timbered, sixteenth-century building and, as it was advertised as a restaurant, we thought to treat ourselves. But directions were less than explicit and the distance further than expected.

When we located the house there were few signs of habitation, let alone a thriving catering business. Doors were locked, curtains closed. We knocked, walked around to the side, and knocked again, until eventually an upstairs window opened

'Clear off — this is private property,' shouted a woman.

Idiotically we apologised, saying we had been told it was a restaurant. Once again the voice floated down to us, 'D'you think I'm opening just for you two? You have to book — and in parties. Now get out'.

The window was slammed and we were left with yearning

stomachs and far from happy thoughts about the Duke and his flaming house.

Luckily Worsley possessed a restaurant to be really proud of—The Casserole, conveniently located a few yards from the canal, near the Court House. Its welcoming atmosphere and the cheerful staff soon caused us to change our minds about Worsley folk.

What an effort it took to leave that comfortable settee where after a delicious meal we sat over coffee and brandy along with all the other customers so that it was just like one large friendly house-party. As people left and bade us all goodnight, we felt we had known them for years.

The bright orange colour in the canal is ever a talking point with boaters and although it is generally known that the cause had to do with minerals in the soil, not many people realise that it is due to ochre emanating from the mines below the canal.

As we travelled along next morning the colour gradually faded until Leigh where the water had resumed its normal grey, muddy colour. Here we joined the Leeds and Liverpool Canal; no stop-lock, no junction, just a name change on the map.

The adjacent scenery now altered. We found ourselves high on an embankment, ahead of us a mechanically-operated swing-bridge near Bickershawe Colliery. When we arrived there we had to leave the boat to find the operator as it was impossible to work the bridge by ourselves.

For the next five miles sheer desolation dismayed us on both sides as we looked down from our watery highway on to a desert of flat, grey, dusty soil interspersed with a few weedy shrubs.

Then ahead we spotted a working boat in mid-stream and could not believe our eyes: its crew were dumping soil and waste matter *into* the canal whereas in every other part of the system, a fortune is spent on dredging and removing the muck and mud! But in this area mine workings cause so much subsidence that the canal constantly requires to be built up, which is why pit waste is used.

The factories of Wigan, with their belching chimneys, were almost a welcome relief and we eagerly sought that butt of music-hall jokes, Wigan Pier. It was a good job we did not blink at the wrong moment or we would have missed it: no wonder it became a joke. As it turned out it was no grand structure stretching out across the water, but just a minute canal wharf and basin, a convenient mooring for town shopping which we decided to use.

To me Wigan has always sounded a dreary, depressing place but it certainly isn't on market day! We had a marvellous time among the many stalls; it was good to see a real market again, with fruit, vegetables, flowers, cheese, clothes, china, books, knitting wool . . . almost as good as our own beloved Maidstone's.

The canal at Wigan ends in a T-junction. It is still the Leeds and Liverpool but now you must make up your mind which way to turn. Go to the right and you are on the way to Leeds. Alas, because of the length of our boat we had no choice but perforce turn left and head for Liverpool.

Much has been said and written about the demise of the canals as a profitable form of transport. Railways, of course, were mainly to blame, but if the many canal companies had only stuck to the same measurement for locks, bridges and tunnels they would probably have been able to put up more of a fight. Sadly the great tradition of the English individualist led to some canals having locks 14ft wide and 70ft long; others having the same length but a width of only 7ft; while yet another part of the system had locks 70ft long but 13ft wide. The Leeds and Liverpool from Wigan to Liverpool has locks 70ft long but from Wigan to Leeds they are only 62ft long. As *Bix* is 65ft long we can never go to Leeds. So Liverpool it has to be.

Soon after leaving Wigan the countryside crept back insidiously. We passed through market-garden districts, with acres of lettuces and fields of vegetables. Recalling memories of those 20 years odd of growing all our own food, we envied the availability of such fresh produce, but not the muscle-binding labour that achieves it.

One drawback to the Leeds and Liverpool Canal is the number of swing bridges, all padlocked to combat vandalism. So we had to buy a key from a neighbouring boatyard and at every bridge I had to slow down for Owen to leap ashore, unlock and remove the padlock, unlatch the bridge, open it (which required superhuman effort in most cases) and then, after *Bix* had sailed through, he had to close the bridge, put back the padlock, run to catch up with the boat and climb aboard. Not easy when every bridge is crowded with swarms of children and teenagers, some throwing stones at me and at the boat, and none making it easy for Owen to gain access to the padlock or the gate mechanism. On top of this wind often made the boat difficult to manoeuvre.

Burscough is the first convenient shopping place on this part of the canal. Long past it was an important staging post for the Wigan-Liverpool packet boat and many of its inhabitants today have relatives and ancestors who worked the boats. We were fortunate enough to tape a recording of a conversation with Mrs Emma Vickers of Burscough. At the age of 83 not only does she recall her childhood when she helped her grandfather, grandmother and her father, all with their own boats on the runs from Liverpool with cotton for the mills, but on a melodeon she plays the canal songs and tunes that used to accompany the boatmen's clog dances.

That Saturday night we moored on the Burscough outskirts, and next morning awoke to find three fishermen sitting directly in front of the boat with two more astern of us. They were quite indignant about having to pull in their lines when we started to move off, which I found rather stupid since they had not been there when we moored.

'You won't get far anyway; wait until you get round that bend,' one cried as we nosed gently into mid-stream.

The canal curved just before a bridge and as we rounded we spotted a police car with flashing lights stationed on it; five or six people by the canal side were looking into the water, and beneath the bridge we could see something protruding above the surface.

As we steered towards the bank and prepared to tie up I noticed a policeman coming down the towpath, slowly waving a large blue and white flag in front of him.

'Do you know what this flag means?' he asked us.

I shook my head.

'I told my bloody inspector you wouldn't,' he grinned. 'He said it means an obstruction in the water ahead.'

'We could see something under the bridge so we knew we'd have to stop.'

'It's a car; we've a diver down to see if anyone's in it. Won't be long. There's a boat on the other side with all the gear to lift and tow it out of the way.'

We were delayed only half-an-hour while the diver confirmed that no-one was trapped inside. Obviously it was just another piece of discarded equipment tipped into the canal — loutish common practice all over the country. We had had a settee, armchair, motor bikes, dozens of bicycles and the odd refrigerator to contend with so far. Now we could add to the list our first car.

More surprises lay in store. As we searched for a likely luncheon place we spotted a line of moored boats, their flags and banners fluttering a welcome. Hurrah! There were other folk using this canal after all.

In fact there was quite a number of people about, some promenading the towpath, others sitting on the deck of their boats and waving to us. We had stumbled on a boat rally, albeit a small one, organised by the local boat club.

We were made to feel welcome and were soon being questioned avidly while drinks were handed round. The club aimed to encourage more local people to use the canal and, judging by the numbers of sight-seers walking about and asking questions and obviously enjoying themselves, I think they could deservedly give themselves a pat on the back for a good job well done.

After our lunch break we prepared to move on.

'How far are you going?' they asked and I called back, 'To the end of the canal, just by the locks, then we'll turn and come back later this afternoon.'

We cruised through areas of derelict buildings and factories presenting dirty, untidy backs to the canal, old bomb sites and miles and miles of bleak, broken-brick walls, and since all towpaths are locked to the public, not a soul in sight.

Our sole solace was the realisation that we were passing Aintree Racecourse. By standing on the roof of *Bix* we could actually see the track, and as the canal curved round a bend, I realised why that part of the course is called the Canal Turn.

The water continued filthy and full of rubbish; pathetic remains of dead dogs bursting from two bloated sacks sickened us. To crown the day our propeller then fouled 200 yards of plastic tape used to wrap electricity cables. As we could see that cables were being laid nearby it did not take us long to realise just where the workmen were jettisoning their junk.

When we reached the canal end on the Liverpool outskirts I heaved a sigh of relief. All we had to do now was to turn around, tie up and photograph the four Stanley Locks that connect the canal with the River Mersey, then we could start back and leave these depressing surroundings behind us. Somewhat to our surprise we spotted two young lads fishing these poisonous waters from the top of the locks, a hole in a brick wall showed why they had experienced so little trouble in finding their way to the canal-side.

Photographing the locks took about five minutes. As we were returning to *Bix,* Owen said, 'You left the back door open.' I knew that I hadn't. Even though we were taking pictures less than 200 yards from *Bix* in fact I had closed all windows and padlocked the door. We saw to our horror that a figure was hurriedly leaving the boat and running across the top lock gates carrying something bulky. Owen gave chase while I, sick in my stomach, ran to the boat. The doors drooped askew, their hinges smashed by a jemmy; once inside it did not take long to see that we had been well and truly burgled.

We had left the boat for five minutes only and were never more than a few hundred yards from it but in that short space of time we had lost a television set, record player, tape recorder, cassette, cine-camera, and binoculars. With tears streaming I wandered the length of the boat: how? why? who? — hopelessly unanswerable questions. And where was Owen? He had rushed blindly off after the man we had seen. Was there a gang? Perhaps he was lying injured somewhere?

I was startled to hear footsteps on the stern deck; it was Owen bearing our tape recorder. He was utterly worn out not only with the weight of the recorder but also with the exertion of the chase. The burglar had balanced the record-player on top of the tape recorder and had run along the towpath until he reached a hole in the brick wall. The player, on top, had been grabbed by a pair of hands on the other side of the wall, and the tape recorder left on the towpath as the thief clambered through to make his escape.

There was no point staying where we were, no way could we get out to the road to contact the police. The hole in the wall led to a factory yard which was railed off and gated — apparently no problem to the thieves if it was to us.

Our return journey was sorrowful as we repeatedly wondered just how the thieves knew of our whereabouts and that *Bix* contained so much equipment of a kind not usually carried by the majority of canal boats.

We navigated ten miles or so before we managed to bring the police on board. Thanks to the towpath being officially closed it was impossible to contact them until we reached the small town of Magull.

Questions, statements, more questions — they went on and on, but it was all worthwhile. No, our goods were not returned. As the police officer said, 'They were probably disposed of long before you reached us at Magull. Definitely the work of an organised gang.'

What emerged from the whole ghastly affair was that they certainly did know about us.

It had all started, it appeared, during our lunch-time stay with the boat club rally. First the gang had plenty of time to carry out a recce of the boats. We now realised that all the pieces stolen were from the same side of the boat, which the thieves had seen through the windows. When we had said our goodbyes to the boaters I had yelled out our destination and our plans.

On reflection we had made it just too easy for them.

The two young anglers we had spotted were no longer fishing when we had rushed back to the boat.

'They were the lookouts,' the police told us. 'The gang would have reached the end of the canal long before you. They had time to park the car for a quick getaway, post the lookouts and then simply wait for you to sail into their trap.'

It certainly taught us a lesson and now we not only lock the boat but also pull down all blinds, and we never, never, tell anyone of our plans. Nowadays we are very vague indeed as to where we shall be, irrespective of the length of the journey. And although it sounds a bit like shutting the stable door after the horse has bolted, at the suggestion of the police we have installed a burglar alarm system.

We were helped with the layout of this by the crime detection officer from the Altrincham police station. Our return along the rest of the Leeds and Liverpool did not take long — we were too heavy-hearted to want to see more of it.

Thankfully we were insured but money was no real recompense for the sickening shock of having our home broken into.

We spent the following month retracing our voyage along the Bridgewater, Trent and Mersey and Coventry canals, catching up with our friends we had made along these waterways and also welcoming members of our family and their friends to stay with us.

With their help I gradually recovered from the upsetting episode, but it was a long time before I stopped worrying every time the boat was left unattended.

4

The early summer of 1977 we spent cruising southwards once more via the Oxford Canal and the River Thames, both dear to my heart.

Much criticism, some heated, is aimed at the Thames; the vast number of boats that sail its waters; the expense of mooring or cruising on it. I am the harshest critic of crowds and costs but I still love the river and was happy with our decision to take our boat to Limehouse where we would join the Regent's Canal.

Born near the river at Greenwich, I had spent much of my childhood playing by it, watching the men in the yards, and being taken for long walks on both sides of the river by my barge-builder father and uncle. So I now looked forward to visiting the working part of the river in my own boat. To many this stretch is dirty, noisy and unattractive, but it forever fascinates me. This time I found sadness too, when I looked at the derelict warehouses and empty docks and realised how pitifully small is today's commercial river traffic.

We took out a five-day licence with the Thames Water Authority for our journey from Osney Lock to Teddington; from there onwards no licence is required for the tidal part of the river.

When our family were young we had enjoyed a week's boating holiday on the Thames each year, over a period of some 12 to 14 years. Routes were never planned; we simply picked up the boat at Maidenhead on the Saturday and irrespective of where we were by Tuesday night or Wednesday morning, we would reverse our course and lazily return to base. For the first three years Reading was our furthest point as there were so many places en route to explore. After that, we felt we knew enough of that patch so we then booked boats further up river, thereby slowly learning about the villages, towns, shops, pubs, parks and playgrounds along the entire stretch. Our children discovered the best spots for swimming, exciting walks, and on one occasion, surprisingly, one of the finest cheese shops I have ever come across. Best of all they came to know the lock-keepers who in turn grew to know us and watched the children mature. And yet we never spent more than a week at a time on the river.

A favourite overnight mooring was at Clifton Hampden where inevitably we dined at the Barley Mow. It became so important a tradition in our lives that our third daughter, Linda and her husband Ian honeymooned there, so of course Owen and I stopped there once more for dinner on our way downstream. Much as we would have loved to, we could not linger as we had only taken out a five-day licence. The days passed pleasantly and we reached Teddington at four o'clock on the afternoon of our last day.

Here we had to check with the lock-keeper on tide times. The entrance to the Regent's Canal through the lock at Limehouse Basin is open only for three hours before each high water. We were advised to wait until the following morning. If we left at 9.30 and took it easy we should reach Limehouse about one o'clock, when the lock would start operating. It was also suggested we should telephone the BWB at Limehouse and inform them of our intended arrival. This we duly did and were told that we would probably arrive rather early but that we would find mooring just outside the lock.

I rang Linda, told her of our movements and asked her to arrange for Ian our son-in-law to bring our post to Limehouse. She in turn, told me that the *Daily Mail* were looking for us as they wanted a photograph to go with the article I had recently sent them about our nomadic life afloat and I was pleased that it had been accepted. I telephoned the features editor who hoped one of his staff would contact us later in the day.

The next morning we swapped that part of the river governed by the Thames Water Authority for that under the jurisdiction of the Port of London. Once past Richmond we found ourselves on rather choppy waters, which *Bix* rode beautifully as we sailed majestically past Kew Gardens returning waves to people as we shot under Chiswick bridge. We completed the boat race course in reverse to reach a part of the river where we would have liked to have stopped overnight. It would have been pleasant to moor at Cadogan Pier, Chelsea or perhaps Lambeth Pier, when we could have invited some old friends on board for the evening, or

The author and her husband aboard *Bix* moored alongside at Maidstone Bridge and (**below**) the entrance to the Anderton Lift

A boat being raised by the Anderton Lift and (**below**) a signpost typical of those to be found on the Trent and Mersey Canals

Entering the Barton Swing Aqueduct on the Bridgewater Canal

Even traffic lights on rivers sometimes fail as did this at Country Lock, Reading on the River Kennett

Marple Locks on the Peak Forest Canal

perhaps visited a concert or theatre. Five days previously when we took out our licence at Osney we had telephoned about a stop-over only to be greeted with derisive laughter; apparently several months' notice has to be given even to stay for one night in the centre of London.

Emerging from Chelsea bridge we heard a shout from the embankment. We could just discern a figure waving frantically which I acknowledged, but the figure started shouting again and, still waving, moved along the embankment in line with us.

I remembered in a flash — it must be the *Daily Mail* photographer! Typical Fleet Street thinking . . . as if we could stop just like that! I waved again and shouted out that I hoped he had a telephoto lens as we could not stop. My words must have floated away downstream but as I watched the figure ashore bent down out of view and then reappeared holding aloft a small child with bright golden red hair. Owen and I at once realised that we were looking at our daughter, Marny, and grandson Daniel, from Tunbridge Wells. What on earth were they doing in London? Marny pointed downstream to Cadogan Pier — obviously she wanted to come aboard. We steered for the pier, whether we had permission or not, and we tied up. Marny ran down the gangplank to us, carrying one-year-old Daniel and tugging three-year-old Benjamin behind.

I helped them on board at the same time explaining that we had no right to stop and tie up. Although the whole operation took less than five minutes, an official appeared and asked to see our papers. He was satisfied with our somewhat jumbled explanation and kindly helped us shove off into the now fast-flowing tide.

As we continued our journey Marny told us how Linda had telephoned her about our Thames trip; Marny had thought it would be exciting to bring the children along for part of it. Her husband, Keith, had checked the tides, worked out the approximate time we would reach Cadogan Pier and Marny had arrived just five minutes before us. Keith had waited in the car to see them safely aboard before starting off for Limehouse to await our arrival.

Young Benjamin thoroughly enjoyed sitting in the bows but baby Daniel was soon tucked up fast asleep in our bed where he remained for the next couple of hours.

When we reached Limehouse we had nearly three hours

outside waiting for the lock to open and the 'easy moorings' we had been told about were nowhere to be seen. Keith, who was waiting for us at the top of the lock, managed to discover a ladder by the side of a derelict timber yard and clambered on board *Bix*.

While we were being tossed up and down by the tide a police launch appeared and suggested that the best place to tie up would be to a group of empty barges anchored mid-stream, as they knew of no moorings by the lock. Eventually we managed this manoeuvre but it was no easy task as the barges, empty and high out of the water, towered above us.

While waiting we hooted several times, hoping that someone would appear at the lock gates as according to the Teddington lock-keeper and the river police, it was well past the lock's opening time. But Keith warned us he had seen no sign of life when parking his car outside the basin.

All afternoon *Bix* thumped and banged against the barges which in turn knocked noisily against each other. We began to worry that the flood tide would soon turn to an ebb leaving us with a three-hour cruise retracing our steps back to Brentford.

It was only by sheer good luck that the lock eventually opened for us.

When Ian arrived with our mail, there was no sign of *Bix* in the basin so he checked with the BWB office. The staff there had no idea of our whereabouts, knew nothing of a boat called *Bix*. Ian decided to walk to the lock and look up river for us whereupon he immediately spotted us swinging against the barges below and tipped off the lock-keeper. A few minutes later we were safely inside the basin — no thanks to the BWB.

I cooked a large meal for all and out came a couple of bottles of wine. Only then were we able to relax. Marny and the children had now experienced the other side of our boating life. We are normally alone when things go wrong which explains the remarks we invariably receive from visitors such as, 'what a lovely peaceful life you lead'.

5

The River Thames of course has played a major role in the transport of goods and people, especially in Tudor times as the palaces they built alongside the river at Greenwich and Richmond bear testimony. But its neighbour, the River Lee, sometimes spelt Lea, was also navigable long before Elizabeth I came to the throne. The Romans, in fact, made considerable use of it, and now the Bryces were about to discover the delights of Izaac Walton's 'Beloved River Lea' — all 27 miles of it from Limehouse to Hertford.

The Lee has been canalised in several parts and the first 12 miles are still used by commercial craft. The transport of timber is the main trade but copper is still carried as far as Brimsdown. We met our first tug, towing four empty barges down river, when we reached Tottenham. I enjoyed seeing working boats again and as we made our way slowly through Hackney, Walthamstow and Tottenham we received many friendly waves from them; when we met at the electrically operated locks there was always time for a chat.

East London waterways sound unattractive for leisurely boating but they are full of surprises. The Lee has supplied a substantial part of London's water since the seventeenth century. Today the thirteen reservoirs in the Lee Valley attract not only fishermen and sailing dinghies but extensive varieties of birds, ducks and swans together with a long-established heronry. Naturalists have counted some 7,000 ducks wintering on the various expanses of water. Not far from Feilde's Weir Lock lies a water meadow which has been preserved in its original state for the last 600 years containing the same species of wild flowers, grasses, sedges and reeds that have always been found in this unusual haven.

From Bow Locks to Tottenham, many preserved old buildings with historical backgrounds are in use. In fact it was the tremendous range of activity on all parts of the river that impressed us so much. And Hackney must be a footballer's paradise for it has an incredible number of pitches on the east bank. My guide book says there are 110!

Every day we met either boaters or bird-watchers, fishermen, canoeists, tow-path walkers, joggers, cyclists. Every inch of water and grass in these parts appears used to its fullest capacity and the Lee Valley Regional Park Authority must be delighted with their decision to make this area the country's first regional park covering, as it does, several miles from London's East End to the little town of Ware in Hertfordshire.

From Stonebridge, just past Tottenham, the locks are manually operated and, with the exception of Pickett's Lock between Edmonton and Ponders End, you must work your own boat through.

A slow and methodical wave from the keeper beckoned us into Pickett's Lock and Owen jumped ashore to help close the gates which, lacking balance beams, are extremely heavy. The same rope, poles and sheer brute force required to open and shut them also applies to the locks at Enfield and Rammey Marsh.

As *Bix* rose slowly in the lock the face of the lock-keeper looked down on me as I stood at the tiller.

'The lady steers her boat called *Bix*
And doesn't get into a fix!'

he declaimed; the first of many instant couplets we were to hear from Harry Mottram, the poet lock-keeper. I like to think that Harry now regards us as friends after four years of using his lock. He invariably accompanies my ups and downs with appropriate poems, some quite long, some beautiful, but mostly guaranteed to get you through the lock with a smile. Harry has published a number of his poems and is one of the waterways' many characters we so enjoy meeting. We feel greatly privileged by being accepted as part of *their* waterway.

Harry was to make me smile on a later visit as I steered into his lock nursing a painful elbow,

'Here comes Iris with her rheumaticky arm
But to me she still has charm.'

His verses may never be called the stuff of literature but they certainly bring joy into our daily routine. By rights Harry should be miserable and crotchety for his is the only manual lock used by commercial craft. Where other keepers press buttons, he has to push, pull and heave and by no means can he be called a young man.

At Enfield Lock we met Bill and Cissy McKay. Bill, retired from river work, had been allowed to stay on in his little cottage which would not have looked out of place alongside a quiet country stream. However, it backs on to the Royal Small Arms factory, the home of the Lee-Enfield rifle, where the sound of gunfire is heard daily, although the plant today only tests weapons rather than making them.

We wanted to leave *Bix* for a week-end while we lectured on canals at a course organised by Theobalds Park College, a couple of miles from Enfield. We had arranged to stop at the BWB Maintenance Yard at Enfield Lock but when the superintendent realised that on the Sunday morning we hoped to bring students from the course to look over our boat and the lock he refused. He could not allow anyone to come through the yard after it had been officially closed on the Friday night.

A pretty predicament! No other boat-yards were nearby and we had just two hours to find a safe mooring and ready ourselves to be picked up by the college principal. So we decided to ask at the lock-house if they would mind having our boat tied up near them, and could they perhaps keep an eye on it during the week-end? The rotund figure of Bill McKay answered the door and his eyes lit up with pleasure immediately he saw *Bix*. It seemed no time at all before we were through the lock, moored beneath his kitchen window and he was on board having a cup of tea with us while explaining that his wife was shopping. He was certain that she would not mind having the boat as a transient neighbour.

Bill chatted about life on the river while we packed our cases and put together our notes, slides and cine-film for our lectures. Our pleasant talk was suddenly interrupted by a cross voice demanding what we were thinking of tying up a boat to a private house. Mrs McKay had returned! Bill promptly vanished to reappear with a surprised but still suspicious white-haired lady. She accepted a cup of tea, looked around, and her white curls nodded, if somewhat grudgingly, as she looked at our home, but clearly she was still unhappy about the arrangement. Who were these strangers appearing out of nowhere like this? Her eyes scanned our bookshelves.

'Why have you that book about Greenwich?' she demanded,

pointing to my copy of *Grandfather's Greenwich*. I told her I had been born there and that my family were river people, barge builders, lightermen.

'What was your name?' she interrupted.

'Challis,' I replied.

'Frank Challis's girl?' She looked hard at me.

'Why yes.'

It was my turn to be surprised.

'I lived next door to your Aunt Aggie and Uncle Fred in Azof Street.'

Now she was smiling.

'Why we was friends for years, weren't we Bill?'

From then on the talk went on remorselessly about those long ago days before World War II. Now it was our turn to worry. Time was marching on and we really would have to leave soon. Hesitantly we asked about *Bix*.

'Well, what about it,' she bristled. 'Leave the boat where it is as long as you like!'

Bill and Cissy, their son and daughter-in-law and granddaughter turned the River Lee into what was almost a second home for us — great river people; proud, generous, hard working with a steady, sensible approach to life that is traditionally handed down from generation to generation.

After our week-end course we moved on. Soon the river abandoned its housing estates and factories; the scenery changed to an expanse of waste land which accompanied us to Waltham Abbey. Here you can moor and visit the Abbey, chosen by King Harold as a centre for religious instruction. Many other historic buildings can be seen in the town itself. A shopping centre, which possesses a launderette, fringes the mooring. It became a regular overnight call on our subsequent journeys up and down river. We had another reason for stopping: to our delight we discovered that the sister of a friend lived there. And what a superb cook Marjorie Brown turned out to be . . . now Waltham Abbey not only means happy memories of boating but delicious meals of jugged hare or fresh salmon poached to perfection!

Six miles further up river is Broxbourne chosen by the Lee Valley Regional Park Authority to be developed as a boating centre. A yard owned by the authority not only builds and maintains boats but hires them out during the summer and provides a daily service by boat from Broxbourne which appeals to school parties and firms' annual outings. The next five miles of river include three boatyards and at least two boat clubs. As we cruised past Cheshunt we could see through the trees beyond the sails of many dinghies racing on the old gravel pits now converted to recreation centres.

Soon we were through Feilde's Weir Lock, at the junction of the River Lee and the River Stort. We continued on the Lee and stopped for the night at Stanstead Abbots, opposite a terrace of pretty cottages. Next morning we visited Ware and were surprised to find that Ware Lock belongs to the Metropolitan Water Board and not the BWB.

At Hertford navigation ends. Here there is a pleasant pub called the Old Barge with the town only a few minutes' walk away. The county town of Hertfordshire is still a quiet, attractive place to explore. So far there is no shopping precinct but I spent a pleasant half-hour in the friendly atmosphere of an old-fashioned arcade of stalls and shops.

The Lee so fascinated us that we decided to retrace our tracks, this time to join the Regent's Canal via the Hertford Union Canal, a one-and-a-quarter mile stretch with three locks and a very obliging lock-keeper who helped work *Bix* through with Owen.

Although the canal is bordered by factories for most of the way the Victoria Park, in Bethnal Green, runs alongside for more than a mile. This is well laid out with exquisite flower gardens and a large lake and attracts many East-enders either to jog, feed the large herd of deer, fish or simply stroll.

A mooring overnight at Old Ford Locks, the first on the Regent's Canal after leaving the Hertford Union, is convenient for shopping so next morning we visited the market in Roman Road, Bethnal Green. Afterwards we returned along the Hertford Union, sometimes called Duckett's after Sir George Duckett who built this link between the Lee and Regent's Canal in 1830.

We now planned to return up the Lee again but this time only as far as Feilde's Weir Lock and the River Stort. Our daughter Rebecca joined us at Broxbourne for a week's holiday; I was sorry she had missed our poet lock-keeper, but a surprise awaited us at Feilde's Weir.

Bix, of course, had navigated this lock before but on this occasion as I waited outside and watched Owen and Rebecca wrestling with the heavy gates, there emerged from the lock house a stocky, bearded man who walked with a limp. I was startled to see him approach Owen with his fist raised. What could be wrong? The lock gates opened and, in trepidation, I sailed in slowly. The bearded figure leaned over and laughed. That raised fist was waving a greeting to us!

'They told me a boat called *Bix* had been through my lock — I knew it had to be you pair of buggers. Oh my, it's lovely, it's lovely . . . quick come inside and have a drink; we'll play some records and . . .'

Don't ask me how we got the boat up and out of the lock; I remember only staring in disbelief as I recognised Danny Ford, an alto-sax player from our early jazz band days, way back in the late forties and fifties.

Naturally we were only too delighted to stay the night and eat with Danny, his wife, daughter, son-in-law, granddaughter — we just had to meet them all — and of course we played jazz records and talked of the old days. What a start to Rebecca's holiday! The weather was glorious for the whole week. Not once were we forced to wear a jacket by day even though it was late October.

Rebecca had to be returned to Broxbourne where she had left her car, but we found time to cruise the first few miles of the Stort, enough to whet our appetite for a speedy return, after we had farewelled our daughter.

The last night of her holiday was spent with Danny. On this occasion we shared our meal with about ten teenagers who used Danny's house as a meeting place where they learned about photography and filming, for our friend not only knew about waterways and music, but for many years had also worked as a cine-film technician.

A happy end to our reunion with Rebecca and Danny.

6

October usually produces a few warm, bright days sometimes sufficient to constitute an Indian Summer. This was the case when we first travelled the Stort. From the moment we left the River Lee at Feilde's Weir the environment changed. Within a few yards we had turned our backs on power station and furniture factory and were cruising alongside an emerald hill where some of the fattest sheep I have ever seen grazed peacefully, or outstared us as we drifted past. Sheep-eyed indeed!

Ahead of us lay 14 miles of quiet countryside with 15 locks to lead us through Roydon, Harlow, Sawbridgeworth and Bishop's Stortford.

In the 18th century, Sir George Duckett had hoped to build a canal from Bishop's Stortford to Cambridge, thereby linking with the Fens, the Grand Union Canal and the north of England. Sir George *did* make the Stort navigable and, when this was joined to the Lee, he said that 'Bishop's Stortford is now open to the world'. Alas, it was not. The other end never did reach Cambridge and today you can boat only as far as Bishop's Stortford itself.

The Stort, however, is ideal for boating, winding and twisting as it does past farms, parklands and country estates, with a few timber yards and maltsters' buildings as a scenic contrast.

The water mills alongside some of the locks are another attraction, although the first, at Roydon, is beside the railway and has a large caravan site surrounding it. *Bix* only just scraped under Roydon Railway Bridge. Before attempting to pass beneath it the water levels must be checked as the river can rise well over a foot after heavy rain.

Hunsdon Mill, one and a half miles further up river is straight out of a Victorian landscape, the adjacent house setting off the mill, garden and river.

Parndon Mill came next, followed by Burnt Mill Lock where

river and railway run side by side and where no mill is to be seen here — hence the name. The railway station also runs alongside the river but don't be misled by its name Harlow Town for it is really Harlow New Town, one of the first communities built to house London's overspill but situated two miles from a part of the river where we were, by no means convenient for shopping.

To confuse matters more, two miles further up river a second lock is called Harlow Lock! Here you should leave the boat as the original Harlow is near and well worth the few minutes' walk to reach it. Harlow Mill, at the lockside, is a restaurant and by the look of the menu is one for that special occasion.

The next town we came to was Sawbridgeworth, and although it is situated away from the river and on a hill, we had a pleasant hour walking around noting its many specialist shops selling such things as hand-made clothes and originally designed toys. The town also boasts two potteries. That part of the town leading to the river consists mostly of new housing estates but we did notice a development at the lock where an old mill, or granary, was being converted into flats, while alongside the lock itself some eight or nine small cottages were taking shape. The opposite side of the river is graced with more maltsters' works and a neat and tidy white timber-faced factory, so altogether this stretch of water is easy on the boater's eye.

The best kept and prettiest of all the mills we saw was at Little Hallingbury. Today it is a boatyard owned by John Wilkinson, an enthusiast, who has restored it to full working order. It is open for visits by arrangement. Mr Wilkinson also has a boat available for river outings.

The Stort enchanted us and despite the coolness of the days the sun shone most of the time. We reached Spelbrook Lock by three o'clock and, although it was only another couple of miles to the end of the river, we decided to stay there for the night. Spelbrook has now lost much of its attraction, principally because the original lock-house has been demolished. As we had seen pictures of this picturesque house with its lovely garden full of old country flowers, we could only wonder yet again what sort of person could have ordered its destruction.

We decided to moor below the lock where we could just see a small house tucked away beside a water meadow. I jumped on to the towpath with the rope as Owen swung the tiller to bring the stern alongside the bank. He then stepped off with the stern rope and we pulled *Bix* into her mooring position. As my rope moved across the grassy bank it snagged on a small mound, removing some of the top soil and dried leaves that covered it. On returning to the boat I was surprised to see a wasp for it was now almost dusk and turning rather chilly. Before I could board the boat I heard Owen shouting and turned to see him with a dark mass circling his head. It was a circle of wasps and they were heading my way too.

I had taken off the top of a wasps' nest with my rope! Waving madly to shake off the wasps we dashed inside the boat for safety, but as we closed the door behind us I heard buzzing from the galley. They were entering through the ventilators above the gas outlets and already the boat seemed full of them. If we opened the windows to let them out, more would enter. I tried opening one window only, and stood beside it to stop further invasion, while Owen tried to drive out the others. It was almost dark before I was satisfied that we had the boat to ourselves again — miraculously we had escaped being stung but I searched meticulously before clambering into bed!

As next day was November 1 we wondered if Bishop's Stortford would be a suitable place to look for temporary work during the winter. We felt disinclined to move too far as both of us wanted to explore thoroughly this beautiful waterway.

Routine took over. True there was not the financial pressure that had controlled our day-to-day existence prior to *Bix* but, as winter approached, it was clear that we should not touch our invested capital — the 'fortune' left after we had sold our farm and bought the boat. If we were forced to sell the boat and live ashore we would have difficulty in finding somewhere to suit both pocket and desire. The sort of house we might afford could not offer what *Bix* gave us.

Each winter, then, we looked for work; so far we had found jobs relatively easily.

During the winter of 1976/77 I had worked in a Gloucester bakery and Owen in an office. The following winter we had worked together in a boatyard. What now? Would Bishop's Stortford have anything to offer? After all it was only a small market town, although it had an industrial estate on its fringes. There was also a good railway service to nearby Harlow New Town which had a number of factories and a large shopping centre. If it came to it, we could probably moor near Harlow if we found work there.

South Mill Lock is the last lock on the River Stort and it was sad for us to see its lock-house and garden derelict. Only a few of the original lock cottages remain and South Mill is one such although we noticed it lacked Sir George Duckett's badge (a red hand) over the doorway as seen on Sheering Mill Lock cottage.

Passing through thc lock, we spotted movement from the other side of the hedge through a gap in which a man emerged. His 'good morning' had a strong foreign accent. We acknowledged, then watched him cross the lock and traverse the next field, wondering whence he had come and whither he was bound.

We sailed on past a timber yard, the backs of offices, a gas board depot, to tie up in the middle of the Bishop's Stortford shopping centre adjacent to Sainsbury's car park.

Our decision to remain there was the start of a wonderful three months. I immediately fell in love with Bishop's Stortford and for the second time only, in all our travels, felt I was glad to be part of this community, if only briefly.

Within a week Owen had secured a job as a labourer in a steel works, sweeping floors and tidying the stores. But long before Christmas he was driving a small lorry, collecting and delivering materials, and working mainly as a stores assistant. We were both invited to the firm's Christmas dinner-dance and not only did I meet his workmates, but became good friends with Andy, the foreman, and his wife Peggy, a friendship that has lasted ever since.

And me? What was I doing? I had found work in a small family printers, a job that taught me a great deal. In addition to printing such items as leaflets, letter-headings and invoices for local firms they offered an ancillary service on books, magazines and encyclopaedias for larger printing works.

Most of the time I stood at a bench alongside the boss's wife and daughter who, among other things, showed me how to rip out a wrongly-printed page from a large encyclopaedia. The thought of handling such expensive-looking books, let alone tear pages out of them, terrified me but soon I was able to work through a crate of 500 and not leave the tiniest scrap of unwanted page behind. Then I learned the art of tipping-in, that is replacing the torn-out page with a corrected one. How my hand trembled as I inserted the first half dozen, but once again I soon acquired the knack. I enjoyed this work more than stapling together sections of the puzzle magazines we printed monthly. Three of us sat at a bench for this operation, possibly the only work we performed sitting down. The magazine, already printed in four sections, came to us in boxes. My job was to place one section inside another, put them on a moving rack which transported them to fellow worker Ruth who combined them with two more sections. When the four sections reached Chantal, the boss's daughter at the end of the rack, she added the cover and stapled the magazine together. As we used up our sections, we would call the apprentice to bring further supplies to keep the machine running. Sections were named by the picture showing on the top page. I was on 'Elephants and chimneys', Ruth on 'Clowns' and throughout the proceedings shouts could be heard 'More elephants . . . another lot of clowns . . . Oh, I've run out of chimneys'. Then the machine would have to stop, while we restocked.

Often it broke down, jammed with staples. I found this frustrating as it usually happened when I was in full swing. However, I enjoyed developing new skills, no more so than on one occasion when we had a rush job which entailed making cartons for sex films, illustrated with stills from the films. As we approached Christmas we started tipping in pages for diaries and then packing them. The firm printed a standard diary and our job was to tip in an appropriate front page bearing the name of the company issuing the diaries.

Oh yes, I certainly enjoyed working in Bishop's Stortford. My colleagues were friendly and although I often found the work tiring, as I could sit down only during my half-hour luncheon break, the experience was rewarding in more ways than one. But back to the boat!

From Mondays to Fridays we moored near South Mill Lock or the Tanners Arms pub a few yards away and on Saturdays we moved to the spot near Sainsbury's car park for our weekly shopping spree. As we wheeled our laden trolley through the car park to the boat a few days before Christmas, a car driver called out, 'That's the way to do it', as he sadly contemplated the long queue ahead seeking a spot to park.

After shopping, we would turn the boat around and cruise down river. Most week-ends our destination was Hallingbury Mill, where we bought gas, emptied our Elsan toilet and caught up with the boating news from John Wilkinson, the owner.

The first time we did this we had something of a surprise at South Mill Lock. Again we were greeted by our foreign friend popping through the hedge, this time inviting us in to meet his wife. 'In' turned out to be a red lorry which served as their home. At the top of steps leading to a back door was Daphne whose cockney tongue contrasted strongly with that of her German husband, Gounod.

Inside we were soon enjoying cups of steaming coffee along with some of Daphne's delicious pastries. Incredible as it sounds the couple had lived in the lorry for several years, travelling through Europe to wherever Gounod's work as a restorer of historical buildings took him. Now they were back in England and hoping to settle in Bishop's Stortford.

The lorry interior was warm and cosy but I would not fancy having to live in it permanently. It seemed rather cramped with four of us in it, but somehow we found room for yet another visitor. This turned out to be Norman, a friend of Danny, the lock-keeper at Feilde's Weir, who had asked him to keep an eye on us and let him know how we fared during the winter.

It warmed our hearts to appreciate that people cared and of course as we continued to visit the lorry every week-end Danny had a regular report from Norman on our activities throughout the next couple of months.

During our stay we had to contend with snow, ice and freezing fog. At one time the water level fell so much that we tilted over at such an angle that we spilled and lost a lot of diesel. Yet, despite the hazards, we were sorry when the time came for us to leave the River Stort.

Goodbyes said to our workmates at the end of January, we started off down river. Our trip was far from the idyllic, lazy journey we had enjoyed on the way up, as the Stort was in flood and our boat rocked along at a furious pace. It took us an hour to work through Harlow Town Lock, and that was with the help of the lock-keeper. He came to Parndon Lock with us, where we found that so much rubbish and broken branches had come down with the flood water that the lock gates were jammed. It was extremely difficult trying to slow the boat as we approached the locks. Each day we faced more obstacles caused by the flood water and, as we grew wetter and colder, I only hoped that things would improve when we reached the wider waters of the

River Lee. At least they had overflow weirs to remove excess water and maintain the river at normal levels.

When eventually we reached the Lee we iced up mid-stream at Cheshunt. After lunch we had to break our way free with boat-pole and muscle. This occurred again at Enfield but this time the ice was so thick that we had to remain for two days. During our five months in the north at Anderton, we had iced up for a couple of days only and that had parted easily as *Bix* cut through it. Yet here, just a few miles from London, we were finding weather conditions more suitable to the Arctic.

One advantage of staying at Enfield was that we could see more of our friends the McKays who lived at Enfield Lock. We told them of our plans to go back up the Thames via Limehouse and their son Don, who was working for the BWB on a dredger at Limehouse, worked out the best time for us to make our trip and kindly arranged to see us into the basin before passing through the lock.

February 21, my birthday, found me back on the Stort — this time by car. A surprise dinner party had been arranged by Owen in that lovely old Mill House restaurant at Harlow Lock, with Andy and Peggy as guests. What a lovely way to say au revoir to what has become one of my favourite waterways.

7

On entering Limehouse Basin my first reaction was one of awe. It was enormous, as were three boats berthed there. When *Bix* left the cosy, intimate, narrow confines of the Regent's Canal and cruised into this vast expanse of water she looked like a youngster's model boat let loose among the real thing.

When we tried to tie up, we realised that this part of the canal system had not been built for boats of our size: we were a good 12ft below the top of the basin and its mooring rings. However, we spotted a ladder on the basin wall and a ring halfway down to which we tied the bow rope. The stern rope was secured to a rung of the ladder with a spare foot or so of rope, a precaution against the wash of any of our monster neighbours hitting us should they sail.

An evening sky of soft orange hung over Limehouse against which were starkly etched the towers of Limehouse Church and an adjoining convent. The silence that prevailed was a pleasant surprise; could we really be only a few yards from the busy East-end of London?

As we ate our dinner in the shadow of a church clock shining pinkly down on us, I remembered one of those ridiculous pieces of information stored in the rag-bag of my mind . . . Limehouse Church clock is reputedly the highest in London. Then the carillon atop the convent tower started to play and the sound drifting across the water transformed our surroundings into an idyll. How contentedly we turned in that night.

The last time I had been wakened suddenly by hearing a loud thud had been me falling out of bed! This time it was the sound of something metallic falling on to the lounge floor. The noise jerked me awake and the next moment nightmare turned into reality . . . the boat started to tilt to one side. Another crash! This time it was our sideboard falling, spewing glasses, bottles, tape recorder and record player on to the floor which by now was almost at the angle where the side of the boat should have been.

Terrified, I scrambled over the sideboard to open our front door and see what had happened, while Owen raced to the stern. A swirling mass of water confronted me under which the mooring rope was fast disappearing. I unfastened it from the ring, thanking my guardian angel that I had used the correct knot when securing. Owen, however, could not reach the rung which held the stern line as it was too far below! He hurriedly clambered back inside the boat and grabbed our carving knife and quickly slashed the rope. The boat promptly recovered its equilibrium and, released from its restraints, rose until we were floating high enough to see over the top of the basin. So dark was it that we were unable to see the position of the other boats. With trembling fingers we tied up again, this time to the rings on top of the basin and then we surveyed the damage, wondering what on earth had happened. Delayed shock hit me with the realisation that if the vegetable colander on the draining board had not rolled on to the floor with a metallic bang, so disturbing my sleep, we could have been drowned as the boat turned turtle, pinning us to the basin wall by our ropes.

Damage to glassware and even to the electrical equipment was minimal, but as we sat drinking a stiff brandy, still shaking with fright, I could think only of what might have been. By now it was three o'clock in the morning and, after much persuasion by Owen, I agreed that there was little point in sitting up reliving the near disaster, so we returned to bed.

Twenty minutes later and still wide awake, I just *had* to recheck the ropes. It was fortunate I did for the water level was dropping fast; already two feet of wall was above *Bix*! A quick shout brought Owen to the stern of the boat. Within half an hour we were 12ft down and tied to the ring and ladder once more, at the full extent of our ropes. We slept that night by fits and starts, my ears alert for the slightest strain on our ropes.

Over breakfast we were still puzzled by what had gone wrong, as the basin was not tidal and a lock lay between it and the Thames.

About nine o'clock our friend Don McKay arrived to whom we reported the night's drama. He, too, looked puzzled and claimed that it was impossible; the water level changed now and then, but never to that extent. He reported our experience to the office and five minutes later was back to explain that we were not the only ones endangered. A particularly high tide on the Thames had swirled over the top of the lock and had been within a few inches of flooding London!

I did not look forward to travelling up river that morning but Father Thames smilingly changed his mood and after our son-in-law Ian joined us for the trip to Brentford in fact we enjoyed a splendid day. London from the river was enthralling and we felt like film stars as we dropped Ian off at Tower Pier to photograph us passing beneath Tower Bridge.

By evening we were safely back via Brentford on the Grand Union whence we joined the Oxford Canal. Two schoolteacher friends from Holland were due to holiday with us for a week and I had reckoned on Oxford being a good pick up point. Our journey there was pleasant as we know the stretches well and had so many friends en route that it felt as if we were returning home.

We arranged to stay in Oxford for Lizbeth and Boudwin's first day and night so that they could 'do' the colleges and their magnificent gardens. As Owen and I had lectures to prepare, we gave them a map and guide books, at the same time telling them we had booked a table at a 14th-century inn for dinner at 7.30. Would they kindly be back aboard an hour before. They left at two o'clock only to return an hour or so later, staggering beneath armfuls of parcels.

I had read that our continental neighbours often came to England to stock up with clothes from a certain well-known chain store, so much cheaper than they could buy in their own country. Imagine my astonishment as our Dutch friends carefully lowered a cardboard-wrapped Flymo lawn mower into the boat, followed by books, pictures and clothes.

They were delighted with the city and thought I was clever to choose a place with shops that equalled London's! We attempted to stow their goodies away in places that would not hinder access to the lavatory, shower and other parts of the boat. Have you any idea what it is like to store a Flymo in a room a mere 6ft 10ins wide? We were soon doubled up with laughter: clearly I was in for a week of surprises with these two. Somehow I managed to persuade them to refrain from buying a second mower for Lizbeth's brother!

Unhappily, our English spring was on its worst behaviour as

we cruised to Banbury through the beautiful Oxfordshire countryside curtained by lashing rain with a gale-force wind screaming through the bridges as we rocked crazily from side to side. Incredible as it may sound our visitors thoroughly enjoyed their first English boating holiday, so much so that the following year they returned with their children and friends to hire their own boats for a week.

On our own once more, we decided to visit the River Nene, waters new to us, although we had boated down the Northampton Arm of the Grand Union to where the Nene started. On that occasion we had moored in Becket's Park, just outside the first of the Nene Locks. These strange-looking locks make the Nene notorious, as do the Hatton Locks on the Grand Union Canal. Not that there is any other similarity. The Hatton Locks are the normal, gated type found throughout the system, all 21 of them! It is the construction of the Nene Locks that makes people say, 'Ah, just wait till you do the Nene Locks, *then* you'll know what it's all about — they're bloody terrors.' The crew of one was about to face those terrors, and as it turned out, I had a few nasty moments, too.

We obtained a lock key from a house near the first lock, and bought W S Williamson's map of the river with some useful details about footpaths, bus stations and pubs. A licence from the Anglian Water Authority is required to navigate the River Nene and when this arrived it was accompanied by duplicated notes on boating the river to Peterborough, a distance of 60 miles.

We sailed a further five miles to the Dog-in-Doublet Lock, the last before the Nene enters tidal waters leading to the sea. The dreaded locks total 37 in number, each 83ft 6ins long and 15ft wide, with wooden gates at the upstream and vertical steel gates at the lower ends. All must be left with their guillotine-type steel gates raised, so that each time you enter a lock on the way down river you have to shut the gate, fill the lock through the paddles on the top gates, then open the guillotine to get out. Coming back you must leave the gate in the up position. As the average number of turns of the lock key required to raise or lower the gate is 96 you can quickly grasp where the hard work comes in.

Almost the first instruction printed in red on our map was a surprising 'Avoid hot water outfall', which we encountered about a quarter of a mile after the first lock. We also noted that care was needed at weirs, locks and sluices if there was a significant flow of water on the river. Our first few locks passed smoothly enough although, once through the locks, the river's strong current made it difficult to stop at a convenient place for Owen to clamber back aboard. The following two locks turned into a nightmare, as the paddles on the top gates were padlocked, partly open. This meant that, as I stood at the tiller, a miniature Niagara streamed into the lock behind me, causing the boat to heave violently in the bubbling cataract.

Luckily we had the company of another narrow boat, *Rose of Sharon* owned by Dr David Owen, an experienced boater and well-known canal writer. After we had discovered two more locks with the paddles in a similar position, he telephoned the water authority to ask why they were padlocked as the river was not in flood and there seemed no reason for water to be run off constantly.

They agreed.

'Their man must have forgotten to unlock them after the previous week's high levels, and they would send someone out,' they said.

They did just that and the rest of the journey was marred only by both of our boats grounding on a sandbank outside Upper Wellingborough Lock. *Bix* seemed fated to spend the rest of her life on these local Goodwin Sands for we were stuck there for nearly two hours! The lock-keeper suggested we tied our stern line to the guillotine gate and he would open the gate gently to see if the rush of water from the lock washed us off. Alas, it didn't! Eventually he sent for a Land-Rover which towed us off. Dr Owen had a much more hilarious time — he watched the lock-keeper put on his thigh waders, get into the river and dig his boat off!

The only comment from the waterways employees was, 'Don't worry, everyone gets stuck there; been like it for years.'

On our return, we took every precaution to avoid the sandbank but again the current took over and we grounded once more. This time our son Jan and a friend who had joined us for a week's holiday, jumped over the side and eventually managed to push *Bix* off. Unfortunately Jan had landed on a piece of broken glass cutting two of his toes quite badly.

However, there were some parts of the Nene I did enjoy: the

lush water meadows that accompanied us for many miles; and the night we moored alongside a field in which hundreds of pheasants clucked and called to each other. I think Owen regretted abiding by the rules and failing to have his 12-bore shotgun with him. I was thrilled to see my first sandpiper, a bird rare to the canal scene. To our further delight, when we spied a great crested grebe it did *not* immediately dive beneath the surface, only to bob up almost hidden in nearby reeds. This one decided to accompany us, first swimming alongside for a few yards only, then flying ahead for a while before skimming the water and landing by our side again. This went on for nearly a mile. I think he was having a game and just showing that he could make better time than us if he so wished.

Because the river floods frequently most villages and towns are situated well away from it, although we did find shopping in Wellingborough and Peterborough most convenient. The Peterborough council have done a splendid job in providing rings and bollards every few yards alongside their park, although the planners obviously did not do their homework. They used paving stone on three levels leading to the water. Each time the river floods, the water softens the soil beneath the slabs which then sink. Workmen lifted a section to rebuild the soil beneath while we were moored there. Ratepayers deserve better of their civic administrators.

We stopped at Oundle to visit this little town and see its famous school. The river was running too fast for us to get into the mooring near the bridge at Oundle Wharf, so we sought — and were willingly granted — permission to tie up at the Oundle Marina close by. From there it was a mile or so into the narrow streets of this lovely old town.

Another pleasant mooring was that near Fotheringay Castle where we wandered over the green lawns to the remains of the last home of Mary Queen of Scots. Nor can I forget the more mundane joy of discovering a launderette three-quarters of a mile from the river at Islip!

The five miles of river from Peterborough to Dog-in-Doublet is straight, giving one a foretaste of what boating must be like on the fens and drains of our eastern waterways. Unfortunately *Bix* was too long to negotiate a bend outside the lock leading into the fens so we were compelled to turn back to Northampton again, but not before being overtaken by a very fast speed-boat towing a water skier just outside Peterborough. Now *that* was something new to us — and our cameras clicked madly as he turned in front of us and repeated the performance again and again. Whatever his speed it made our three miles per hour seem as if we were standing still.

8

During the summer of 1978 we decided to mix old with new, so apart from travelling much-loved canals *Bix* also ventured on unknown waters. We met old friends; the family holidayed with us; and we had three bookings from paying guests.

Among the boat hirers to whom we talked as we worked through locks or moored nearby were a young couple obviously on their first boating holiday. She was having trouble operating a lock and gladly accepted the help and advice that Owen offered. Later that evening we saw them in the New Inn, at Long Buckby and soon we were chatting away like old friends. Nicki and Richard were both actors taking a spur-of-the-moment holiday during one of those all too familiar periods of their profession known as 'resting' which, to the rest of the world, means out of work!

Richard had spent his childhood in Hemel Hempstead, and having watched the boats on the canal daily had developed a tremendous interest in the waterways although it was all new to Nicki. During that first evening's conversation, however, it was clear that she was already succumbing to the bug.

On the last evening of their week's cruise we invited them on board for coffee and drinks. As they enthusiastically related the adventures of their last few days Owen and I realised how much now these two people were hooked on canalling. We looked at each other, and I nodded in agreement,

'Iris and I have quite a few week-end lectures booked,' Owen told them, 'and as we do not like to leave the boat empty we let our children and friends use it. Would you be interested in looking after *Bix* for us sometimes?'

Their first reaction was of total surprise; Nicki was speechless and Richard stammered a few unfinished sentences most unprofessionally,

'But it's your home . . . you mean you'd let us . . . ?'

'I'd be terrified to move it,' added Nicki for good measure.

We explained that they would have to spend a day with us; Owen would go over the technical side with them and I would show off the domestic arrangements. They agreed with enthusiasm; three years later they are still among those on top of our list as deputy crew, even though their first week-end wasn't exactly a success.

We arranged to hand over *Bix* at Enfield Lock on the Friday morning, meeting at the adjacent Greyhound pub. We would collect the boat two days later on the Sunday evening. Fine. We returned to a tale of startling events beginning with Richard's discovery of a new way to empty the Elsan chemical loo; he dropped it while carrying it through the boat!

Nicki's expression as she described the scene and of how many floorcloths and sponges, along with gallons of disinfectant, she had used to clean the mess should have earned her an Oscar!

'*Bix* certainly has the cleanest floor of any boat on the canal system,' she concluded.

Their saga had continued with Nicki discovering that, unlike canals, rivers do not have a continuous towpath and that it is often difficult to find a suitable place outside locks where one can easily return to the boat. Nicki had fallen in on one occasion and as Richard fished her out of the River Stort it was *his* face this time she said that deserved the Oscar!

The next part of their story was less funny, as they sat contritely round the saloon bar table at the Greyhound. They insisted we had another drink before we left to board *Bix* which they had moored a few yards away.

Having taken our advice they had travelled a short stretch of the Stort to Roydon where they turned before the lock . . . and, at the very moment when you do not want even a zephyr, a gust of wind had swept the boat sideways into an overhanging branch of a tree which had broken the bedroom window. With the help of the Roydon lock-keeper, the hole had been firmly sealed with a bright blue plastic bag.

'But we've found a glazier in Enfield,' they gabbled on, 'and we'll be back in the morning with it all repaired.'

By now we were laughing, knowing only too well how helpless they must have felt when the wind took charge. We, too, had had our disasters; and only three strips of glass needed to be replaced in our louvred windows. All ended well, so well in fact that a few months later when they took *Bix* down the River Wey for us the only thing they lost was our chimney! Still they had encountered the worst gale recorded in Guildford for several years, and at least it made us put a chain on the replacement stack. Our theatrical caretakers certainly sing for their supper when they recount their 'disasters'.

Our first paying guest was Peter King, an interesting character. On a two-months' vacation from Australia he was visiting his family in Hertfordshire from where he had emigrated 30 years earlier. Two weeks of his holiday were to be spent seeing England from a canal boat. We picked him up on the North Oxford Canal at Rugby. He selected the Grand Union route to Norton Junction, down the Leicester Arm on to the River Trent. There we took the Cran Fleet Cut which, in turn, becomes the River Trent, the Beeston Cut and, finally, the Nottingham Canal. The route is as changeable as the names; woods, flooded gravel pits, industrial estates, and finally Nottingham itself.

Our guest was an Arctic explorer, who had brought hundreds of slides of his last expedition to show us after dinner each night. It was certainly unusual entertainment to be looking at icebergs, glaciers and sled-pulling dogs, all at the height of our summer. He also showed us the mountain named after himself — King Mountain. Peter's visit would have seemed like a holiday to us, had it not been for the extra work to provide our guest with the best cuisine and comfort available on a canal boat. I enjoy cooking and entertaining but somehow it always turns into a chore when I have to do it for a living.

We retraced our path at Nottingham as our guest wanted to leave the boat at Burton-on-Trent: I only hope he was as fascinated as ourselves. I shan't forget edging through a secret, silent Nottingham on a Sunday afternoon, past factories built when cheap labour had permitted ornate brickwork even for a workshop window — in some cases a different pattern for each storey of the factory. And I wonder whose sense of humour influenced the notice on the back of the Portland Hotel depicting a uniformed man with a saucy grin, above his head the words, 'Night porter at your service'. It reminded me of the church we had once seen outside which the vicar had displayed something similar: 'I'm at your service. Are you at mine?'

I was glad to have a four-week break before the arrival of our next guests during which time Owen and I attended a jazz summer school while our son and his friends holidayed aboard *Bix*, taking her to Oxford for us. Their log read that it had rained nearly every day, including a cloud-burst or two; we could not help smiling at the final entry: 'Thank you for a lovely holiday.'

We cruised the Oxford canal to the bottom lock at Napton where a message awaited — requesting us to telephone our daughter Marny.

'Where will you be by two o'clock on Saturday?' was Marny's first question when we got through.

'Probably on the Grand Union near Warwick,' I replied.

'Great! Keith and I are going to a wedding in Birmingham. Can we dump the children on you? They'll love sleeping on the boat — and we'll be able to stay on for the wedding party. We'll pick the boys up sometime Sunday — oh and have lunch with you!'

So at five past two on Saturday afternoon I was busy stowing away all the paraphernalia that comes naturally with a four-year-old and his two-year-old brother! Who says we are not normal grandparents when we are available for baby-sitting at a moment's notice? Mind you Marny and Keith could have chosen a better time: there were 46 locks to get through before meeting them, but at least it gave us the incentive to scale the Hatton Flight.

Ben and Dan certainly enjoyed sleeping on the boat and they were so good that I wished their stay had been longer. We remained at our mooring beyond the top lock of Hatton, a convenient spot for the children to be picked up on Sunday. For walks along the towpath, my two grandsons were perfect companions, particularly when we watched a brood of five fluffy black chicks frantically swimming to keep up with mother moorhen. And, of course, out came the two brightly coloured fishing nets we keep on board for such occasions — but try as we might we could not catch enough fish for Owen to have for his dinner! It was well after teatime on the Sunday when two sleepy little boys were tucked into the back of the car for the journey home to Tunbridge Wells. The wedding party had been super; but our week-end had been even more rewarding.

Next morning we were off again to complete a boat trip that had started in 1968. At that time it was possible only to boat the River Avon from Tewkesbury to Evesham. Now, thanks to the skilful organisation and tremendous enthusiasm of David

Hutchings, and the physical work and financial contributions of countless other helpers, the Avon was navigable to Stratford-on-Avon.

Our approach was from the Stratford-on-Avon canal which joins the Grand Union five miles from Lapworth, one of the great names for canal buffs but meaningless to the rest of the world. In the heyday of canals it was an important junction where the village had developed because of the canal.

The stretch of water from the top of Hatton locks to Lapworth was beautiful to cruise on the bright summer's morning that greeted us; we passed through woods, lush green fields and, halfway, a small tunnel at Shrewley. As I steered *Bix* through the fairytale scenery I tried to put myself in the place of the engineers who had planned its course all those years ago. It was as if they had said,

'You think this is lovely? Just wait until you turn the corner at Kingswood.'

There a short stretch of water leads to the meeting place of the northern and southern sections of the Stratford-on-Avon canal, between locks 19 and 20. On the southern part, which was transferred to the National Trust in 1965, we had to buy another licence from an office by the side of a picture postcard lock. How right those engineers were.

Now that the Upper Avon is navigable, it is possible to cruise from Lapworth to the River Severn, for which you need three licences: one for the National Trust's Stratford-on-Avon canal; a second for the Upper Avon from Stratford to Evesham; and a third for the Lower Avon from Evesham to Tewkesbury. The trip cost £24 in licences which successfully removed our spending money and most of the joy from our day.

The countryside is undoubtedly exquisite. The waterway meanders through magnificent Warwickshire scenery graced by the unusual architecture of its canalside houses, giant barrels with rounded roofs so designed by the canal engineers. Their experience was limited to that of digging canals and building tunnels — so they literally built a tunnel above ground and surrounded it with walls.

The extremely narrow bridge holes throughout the Stratford-on-Avon canal lacked towpaths for the horses used in those far off days, so slits were left in the top of the bridges for the towropes to pass through. The horses could then continue towing around the outside of the bridge. Little money was spent on the construction of the canal, still the trouble today. The 35 locks meant hard work for a crew of one: several had no balance beams, and one had a moth-eaten looking railway sleeper hanging precariously from the gate. When we struggled to open the lock it promptly fell off, narrowly missing Owen's foot.

By contrast most of the towpaths are beautifully kept, the people who live by the canal obviously taking great pride in them, with the grass as neat and trim as their own lawns.

The water level for much of the canal is low so we often ran aground even though our draft is only 1ft 9ins. But the quiet rural canal is much more in keeping with Shakespeare's England than the coach-laden roads that carry the hordes of tourists around Stratford. The only sign of industry to be seen in this area echoes the flavour of the canal; it is the boatyard at Wootton Wawen, situated in an embanked basin and sufficiently well designed to earn a Civic Trust award in 1972. The canal is carried over the A34 by an aqueduct whose height is now too low for modern transport causing it to be damaged several times by lorries hitting the underside. This of course holds up all boating and the cost of repairing the damage must be a major drain on National Trust finances for maintaining the canal.

There are two other aqueducts on the canal, a small one over a tributary of the River Alne at Yarningdale, and a larger, the Edstone or Bearley Aqueduct which crosses both a road and railway. A towpath runs alongside at the level of the bottom of the aqueduct, and it is a strange experience to walk beside your boat and look up at the helmsman, the nearest sensation to walking on water that I have had!

We found a pretty mooring at Wilmcote, an attractive canal-side village, the home of Shakespeare's mother, Mary Arden. I wonder how many tourists had our good fortune when we visited Mary's home. We merely had to leave our boat, take a few minutes' walk and there it was — the original 'ye olde' English thatched country cottage complete with old-fashioned garden, and only three other tourists in sight, for we had arrived between coach parties. As we wandered around leisurely it was as if we were visiting an old friend's house. There was no guide to explain its history, just a book to help us with the details that make such a visit so fascinating — and all unhurried in our own good time. Certainly it was a rewarding experience.

With 16 locks in the three miles now between us and Stratford we were glad we had stopped for a delectable lunch in Wilmcote to provide the energy. The locks start immediately on leaving Wilmcote with a flight of 11, followed by a longish pound with one further lock before the last four are reached. Railway, gasholders and the detritus of city life could be seen from Bridge 64, but the canal itself, apart from a few houses and a small block of flats, maintains an isolated privacy of its own.

The last bridge before the canal enters the basin in Stratford is particularly low. You emerge from Stygian darkness into the hustle and bustle of crowds lining the basin, their cameras clicking around the Shakespeare Memorial Theatre, indeed a world of a very different kind.

A few boats were moored in the basin but we found the ideal spot, almost outside the theatre front. As soon as we had tied up I ran to see if there were tickets for that evening's performance. I was about eighth in the queue, and what a civilised way to wait: chatting in the foyer for *Bix* was causing some excitement moored as she was so close to the theatre where my companions' questions about our way of life made time pass quickly. Owen spelled me while I returned to the boat to prepare our evening meal. At 6.15 we swopped places again and I decided to give it another fifteen minutes only. If we could not get tickets we would spend the evening wandering around town. But before my self-imposed time limit had been reached I was back aboard holding two precious tickets in my excited hands for the front row of the dress circle!

It was certainly our lucky night — the perfect ending to our journey down canal. A modern production of *The Taming of the Shrew* used musicians on the stage as part of the plot. When the trumpeter stepped forward for a solo Owen nearly toppled over the edge of the balcony: it was our friend John with whom we had spent the previous Christmas night. During the first interval we left a message for him with the doorkeeper; during the second he invited us to meet him backstage afterwards.

The evening ended with drinks on *Bix*, nearer than any pub and just a minute's walk from John's dressing room.

We arranged to meet for lunch the following day and when John and his girl friend arrived there were two other musicians on board. We had awakened that morning to the sound of a clarinet playing *Memories of you* emanating from another canal boat, *Faraday*, moored opposite. The musician was its owner who had read about our piano-carrying boat so he and his pianist wife came over for a jam session.

To continue our journey it was necessary to buy a licence to enable us to cruise the Upper Avon. But the office had run out of passes so we were instructed to get one when we reached Evesham, and thus unlicenced we proceeded down to the River Avon.

An outstanding view greeted us as we emerged from the Stratford Lock: Holy Trinity Church and Theatre, both reflected in the water. Before reaching Evesham we had eight locks to navigate, each a Heath Robinson affair, some built of concrete blocks, others of steel with gates and wheels given by or recovered from other waterways, including the Thames. Some of the locks were named after people who had given donations for their reconstruction, an intimate touch. Sadly the last one, the George Billington Lock, was completed only four days before Mr Billington's death in December, 1969. Invariably the locks are situated on sharp bends or where obscurant trees hide navigation marks and 'Danger Weir' signs.

With happy memories of our holiday on the Lower Avon in 1968, I had looked forward to this trip, but although weather and scenery were pleasant with thick forests occasionally lining the river's shoreline — much to my liking — there was little joy for us on this occasion, due mainly to lack of mooring facilities. Nothing is more frustrating than to read in your guide book about a delightful building or village only to find that the adjacent riverside for a mile or so is privately owned thus making it impossible to tie up and explore.

We had difficulty in finding somewhere to stop on our first night. Old hands at boating the Avon had warned us that only by arriving early in the afternoon could an overnight mooring be ensured, and by three o'clock a few places on the map were already full. We usually tie up about five o'clock but we were still travelling an hour later than this with not a mooring in sight. By now a steady drizzle had started and curtained us with a cold mist, hiding lock signs until we were almost on top of them (an additional hazard for canal boats with tiller steering).

At a quarter to seven, two dripping, tired disconsolates espied a large grassy bank approaching the lock by Cleeve Prior.

Thankfully we tied up there and half an hour later were taking a well-earned rest while we waited for the meal to cook and accompany the bottle of wine on the table.

My habit when we tie up is to take a walk along the towpath; this way I really get to know the waterway and its environment, especially if we are exploring one particular stretch time and time again. Through this habit we have come to know many water lovers and even recognise some of their dogs and cats! And it makes our day when a local recognises us. Then we seem almost to belong to their community, at least briefly. The difficulty with rivers is that there is no continuous towpath. Somehow or other this has usually been overcome, even on the Thames where many public footpaths have been established. The Upper Avon proved an exception — only on one occasion near Bidford-on-Avon was it easy to walk beside the river. On our first night's stop, the drizzle turned into a downpour so we stayed snug and dry inside *Bix* and enjoyed an early night, promising ourselves a short exploration in the morning.

A low grey sky with more drizzle greeted us when we awoke, but as I prepared breakfast I could see through the window a tempting stretch of grass leading to a wood. What happened next is difficult to explain unless you have sheltered in a corrugated iron shed during a hailstorm.

A volley of crashes on our roof echoed inside *Bix*'s steel frame and I yelled for Owen who ran to the steps from the music room to the stern only to duck quickly back as the clanging on our roof started again, this time accompanied by a stream of foul language. When the barrage eased and it seemed safe for Owen to emerge without having his skull split in two, the mystery was explained.

An elderly man, who had been banging his walking stick on our roof, angrily informed us that we had moored on private land. In turn I became annoyed not only by his attitude but also because he had scratched the length of our newly painted boat!

We asked him where the notices were to indicate the land was private to which he gestured with his walking stick. We repeated our question and he walked through thick undergrowth for 50 yards or more to unhook clinging bramble from a drunken notice board. When I followed him to read it he told us we could be prosecuted for trespass.

This upsetting episode destroyed all savour for breakfast so we retrieved our mooring spikes and moved off with only one thought in mind — to get away from the Upper Avon as expeditiously as possible. A sad tailpiece to all the work instigated by the man who gave heart and soul in restoring this waterway for navigation again — David Hutchings.

At Evesham we bought our Upper Avon licence for which I grudged every penny and it will be long before I cruise that waterway again. At the same time we bought our licence for the Lower Avon, and I looked forward to revisiting some of the many pleasant spots we had come to know when holidaying along its length. First priority was dinner at the Old Vinecroft in Evesham, a pleasure we had to forgo as we simply could find nowhere to moor, the public stretches in Workman Gardens being full. We even had to tie abreast of another boat to replenish our water supply and empty our rubbish. I fail to understand why local councils have not increased mooring facilities now that this river is used by so many boats which surely must bring so much more trade for their shopkeepers.

We were inured to disappointments on the Avon but, happily, the views are still magnificent. Bredon Hill looms up continually on the horizon as the river winds and twists through the Evesham valley; orchards climb neatly and tidily from the river although the fields must be difficult to farm with gradients so steep.

As we had been cheated of our dinner at Evesham, we made for Pershore having happy memories of eating at The Three Tuns there but this lovely town could offer moorings suitable only for small cabin cruisers with nothing available for a long canal boat. By the side of the town's approved mooring site was a notice warning of prosecution for mooring beyond the board. I wonder how the people of Pershore feel about their waterway's 'welcome' to questing travellers.

We didn't bother to find out and headed on to Great Comberton for the night. The next evening we treated ourselves to a sumptuous dinner at the Hop Pole Inn in Tewkesbury, a lovely old hotel with not fairies but yes, you guessed it, moorings at the bottom of their garden.

The beautifully-restored Little Hallingbury Mill on the River Stort

Hunsdon Mill Lock on the River Stort and (**below**) the ghostly Gothersley Lock

'he author's old home territory — Greenwich Royal Naval College from
ìreenwich Pier

uthor and family evidently warming up for one of their regular musical evenings

Bow Creek on the River Lee and (**below**) the Clock Mill, Three Mills, Bow on the Ri
Lee, a conservation area of fine Georgian Industrial buildings

9

If things had gone according to plan we should have had a choice on reaching the River Severn; down river for another cruise on the Gloucester and Sharpness Canal, or up river to visit that gorgeous gift shop in Upton-on-Severn with the fabulous clothes on the first floor. Not that I could afford any of their beautifully-cut suits or cuddly tweed skirts and in any case my life simply did not call for such apparel. As it turned out we had no option when we left Tewkesbury and locked down to the Severn, for two American school teachers had booked a long week-end with us and we were picking them up on the Staffordshire and Worcestershire Canal.

Since selling our farm and joining the world's nomads our life had changed in a way that dreams never foretell. Our two years of idyllic, lazy meanderings around the waterways were to have been an interregnum, a time to stand and stare, to take stock before plunging back into the world to start another chapter in our lives.

But now after being afloat for twice as long as originally planned, we were firmly settled in a lifestyle that not only provided a perfect method of spending our leisure hours, but whose economics meant that we no longer had to work at the grinding pace that previously we had found necessary. It meant, however, that at certain times of the year we had to be in a particular place, either for temporary work or because we had to pick up paying guests such as our American visitors.

We cruised up river to Stourport, by no means a chore even though we did not have time to stop at Upton or Worcester. There is always a tomorrow in this new existence of ours! We called at Severn Stoke to visit a fine trombonist friend who owns the country club there, and where Owen fancied his chances of sitting in with the club's band. One of the few disadvantages of being a nomad is that Owen no longer has his own band, and after more than 30 years of regular playing he has to grab opportunities as they float by!

Surprisingly, he does not appear to miss the old life. I think that this is because the camaraderie between jazz musicians who play together can be appreciated only by those who have experienced it. This type of music has a closely-knit minority following that makes it fairly easy for Owen to rejoin whenever he meets with fellow enthusiasts, wherever we happen to be. So Severn Stoke made a very happy mooring and we had a particularly warm welcome there. I find it always gives me great happiness to see and hear Owen as part of a jazz band again.

Next day we tied up at Stourport, a town born of the canal age in the late 18th century and where our own love of canals was born. On a Severn boating holiday many years ago, we had moored on Stourport's riverside and wandered around the town. Here we discovered its other, and much more fascinating, world of locks and basins bordered by glorious 18th and 19th century architecture. In one corner of an upper basin stands a narrow bridge which is the intriguing entrance to the Staffordshire and Worcestershire Canal.

Susan and Frances, our two guests, were to be picked up at Compton, about 20 miles along the canal. We had plenty of time for an easy trip on our own through Kidderminster and Kinver, and as we worked through the Bratch Locks we decided that if we returned this way we could show off not only some beautiful English scenery, but unusual locks and, of course, the famous Kidderminster carpet factories.

We had arranged car parking for the Americans with a publican in Compton and by three o'clock on Friday *Bix* was resounding to, 'Gee, this is going to be great,' and, 'Oh, my what a cute boat this is,' as Susan discovered the delights of our home. Frances was absorbed with pencil and paper, noting the many wild flowers on the bank. Botany was her special subject and I spent many frantic moments during the next three days fingering my copy of Keble-Martin's *Concise English Flora* so as to be able to give the correct names, English and Latin, to the flowers and plants known only to me by their common nicknames, such as Lords and Ladies, and Shirt Buttons!

We had about 44 locks to work through during the trip down the canal and back to Compton, and after the first both girls were hooked and eager to turn windlasses, and open and close the gates. This eased life for us, as Owen could steer leaving me to prepare meals or sit and relax. It was not often that happens when we cruise!

Bratch Locks came as a surprise to them. These locks resemble a staircase, with one lock leading directly to the next, although in fact they have a short pound between, possibly the shortest in the country. The pretty octagonal toll-house there just makes you grab for a camera.

One afternoon Frances exclaimed joy and admiration as the sun shone on a vast bank of rosebay willow herb, the world refected in its rosy glow. Almost immediately afterwards we passed through Cookley's tiny tunnel and emerged to see by way of contrast a towering cliff overhanging the water: Austcliff Rock, covered with vegetation drooping low, and conveying the impression that the boat cannot possibly continue any further. Somehow or other it slips slowly beneath the awesome red stone boulder whereupon thick woods sweep down to the water's edge.

I had planned a mooring that night at Gothersley in order to relate to our guests a story about the attractive white house by the side of the lock, its garden bordering the canal. The house was empty, the grounds a wilderness where waist-high grass covered one-time flower borders . . . looking exactly as it had two years earlier when Owen and I had first seen it. At that time we had a vague notion that if the right property came along we could be interested in buying. This place was prettily situated in rural countryside, close to the canal, but was not BWB property. We had walked around then, peering through dusty windows at rooms which were precisely our size, not too poky but not so large that their heating would cost a fortune. The garden, too, would have suited us, sufficient space for flowers and vegetables, and an easy shape to work. As I walked the boundaries on that first visit I took a sudden dislike to the house and promptly returned to the boat. When Owen asked what I thought about it I replied that I could never be happy there. And that was that.

Later that night we had moored outside Greensforge Lock, a couple of miles further on, where we met the woman living in the lock-side cottage. I mentioned the empty house, as it is rare to see such property for sale by the canal, and she confirmed that it *was* indeed for sale but no one wanted to buy it. All she knew was that it had been built a few years ago by a young couple from

Wolverhampton who had been happy there for a short time only: unfortunately the woman simply could not settle down to the house, feeling nervous and unhappy there. As it was situated in an isolated spot, they decided to sell up and move back to Wolverhampton. Strangely, the second owners had also stayed for a brief period only when they, too, moved.

'No one knows what's wrong but that house isn't a happy place,' our storyteller continued.

The evening's gossip then turned to gardening and the house was forgotten.

Nearly three months after that conversation we had a dinner party on board where our guests were Derek and Frances Pratt, Derek well known through his canal photography and Frances for her articles and lectures on canals. After dinner Derek told us he had just experienced a nerve-wracking assignment photographing some locks. As it was a lovely sunny day he left the boat and, taking a windlass, decided to walk along the towpath and prepare the next lock.

Having done this, and with time to spare, he walked beyond the lock to photograph the scenery in front of it. As he did so he heard the paddles being wound up and the sound of water running out. He turned to warn whoever was there to leave the lock alone as a boat was approaching, but no one was to be seen! Believing that a child was playing about he walked back to the lock and beyond it, but there was not a soul in sight. He refilled the lock and waited by the gates for his friend's boat to arrive. As they came into view he walked towards them to take a picture of their entering the lock. As he did so, he again heard the sound of paddles being raised and water running out. He ran back, convinced that someone was opening the paddles and then hiding in the bushes. Again he found no one. When the boat reached the lock Derek and his friend worked the paddles in the normal way and went through without any more trouble. . . .

At this point in the story I interrupted Derek.

'I know what lock that was,' I heard myself saying.

'I haven't even told you which canal I was on,' protested Derek.

'It was Gothersley Lock,' I said and outlined my reaction about the house adjoining that lock. Silently we had another brandy apiece; there was no more talk that evening of psychic phenomena.

So here we were back once more at Gothersley as I recounted my story over dinner to the two Americans. Immediately they demanded to see the house, so all four of us walked along the now very much overgrown path to its front door. Nothing had altered, but this time I had no reaction and the building certainly looked beguiling with the late evening sun shining on to its plain white walls but as I looked a thought brought me up shortly. In two years no vandal had touched the place; windows were intact, the walls were free of the usual paint-sprayed graffiti. Even the garden fence sagged only from lack of maintenance, not one paling had been damaged. Why was this place unspoiled and still habitable while every other house, shed, barn or whatever, left empty by the canal-side, is vandalised in the first week it is empty?

Whatever the reason the story made a good finish to our guests' week-end on the canal.

10

Once upon a time a boating holiday took the form of hiring a boat and cruising along a canal or river for three or four days, then retracing one's steps for a similar period, returning to the boatyard in a happy, relaxed frame of mind.

As more canals were made navigable circular trips became more favoured; today if a ring of waterways is not travelled in one or maybe two weeks, then the holiday has been unsuccessful, even if completing the circle means there is no time to stop and savour the life in the places you pass through. Today's holidaymakers consider it a waste of time to boat back along the same stretch of water. If only they would realise that the water world has a different face when seen in reverse, as it were.

These thoughts sprang to mind as we cruised — for the umpteenth time — the Trent and Mersey, Macclesfield and Peak Forest canals, to Whaley Bridge discovering, as we always do, an infinite variety of new experience to surprise us.

I noticed an original ornament in one of the gardens near Alsager — a set of white-painted iron fire-dogs. They did not look at all incongruous on the lawn, away from their usual position by a fire-place. In fact they were the focal point of the garden, in the centre of a grassy slope, and jolly good they looked too.

I was glad to see my favourite shopkeeper in Kidsgrove, the one I have yet to see wearing shoes! In stockinged feet he walks around his shop weighing fruit and vegetables. It is the many pieces of cardboard stuck on till, scales and shelves that makes his shop such a delight. These are covered in hand-printed quotations and home-spun philosophy. As I waited for my carrots to be weighed I read,

> 'They said, "Smile, things could be worse,"
> So I smiled, and things became worse.'

I too smiled as I left the shop; what a relaxed and amusing way to buy the weekly food, one up on the usual supermarket chore.

We stopped overnight near Congleton, on the Macclesfield Canal, and chatted with a sprightly 80-year-old character called Tom who had worked on the boats for more than 40 years. His favourite run had been a 14-day trip which started by taking a load of chemicals from Froghall, on the Caldon Canal, to Gas Street Basin in Birmingham where they were transferred to another company's boats. Brierley Hill had been Tom's next stop; here he took on bricks and delivered them to the kilns near Wheelock. There soda ash or brine was loaded and carried back to Froghall.

Tom had this trip worked out so well that always he managed to be at Planet Lock on the Caldon on a Saturday night, so that he and his mates could go to Hanley Empire, the 'Blood Tub' he called it, to enjoy the music-hall turns.

How the canal people came to life as Tom standing beside *Bix*, his Jack Russell terrier at his feet, told us proudly of his working days. Fancy missing someone like that just to keep to a fixed timetable in order to complete a wretched circle.

At Whaley Bridge, we picked up our last paying guests of the season, and what a pair they turned out to be. Rae and Tony Kozary must be the country's greatest canal enthusiasts. Rae has been holidaying on the canals since 1957 and Tony a few years later. To our surprise they told us they had never handled a boat themselves; always they had had a hotel boat holiday, until they read about us.

We had a fantastic week with them, and we still meet and write regularly to each other. I think *Bix* means almost as much to them now as it does to us. Their holiday consisted of cruising along the Peak Forest Canal to Marple, then down the Maccie to Kidsgrove where we turned and retraced our route to Whaley Bridge. They knew the value of that return trip and the chances it gave Tony to photograph the bridges and locks he missed on the way down!

After saying au revoir to our guests whom we *knew* we should see again we retraced our way south, repeating the previous week's trip to Kidsgrove, through the Harecastle Tunnel, and along the Trent and Mersey.

We then joined the River Soar and made our way to Leicester. The Soar has been canalised in parts and eventually becomes the Leicester Arm of the Grand Union Canal, joining the main section at Norton Junction.

We headed for Leighton Buzzard and a first-ever family Christmas aboard. Rebecca and her two children, Erica and Dylan, were to spend four days with us starting on Christmas Eve. It had been agreed that not only was Leighton Buzzard an attractive part of the canal but, more important, it would be a fairly comfortable drive for Rebecca from Tunbridge Wells. I also knew that the town itself had excellent shops, not only for festive food but for any extra gifts that we needed for our ever-growing family.

We left Whaley Bridge on October 7 allowing ourselves six weeks to journey approximately 205 miles, fully conscious that four hours travelling a day would be our maximum. This would allow a good week for shopping, cooking and, by special request from Rebecca, buying and decorating a Christmas tree! As I steered *Bix*, it was the tree that was constantly on my mind.

Where on earth would we put it?

Our lounge may be 30ft long but it is only 6ft 10ins wide. And this raised another problem. Where would we hang the paper decorations, not to mention holly, ivy (both easy to obtain free of charge when canalling) and of course, balloons for the children?

On the previous three Christmases, which we had celebrated on our own, I had made a token effort by putting some greenery around the pictures; but now, with Rebecca asking for all the traditional bits and pieces that had always been part of our family Christmas, there was much to do. And we had also to think about the children's stockings. How could we hang them up at the end of sleeping bags on the floor?

All this and more took precedence over everyday happenings, such as taking notice of the places we were passing through. I had no time to chat to Owen about the antics of moorhens, or the late roses blooming in canal-side gardens. I steered and pondered about Christmas until Leicester came into view. We spent a day shopping there and enjoyed selecting presents for the grandchildren. The weather was pleasant and even sunny, although we were beginning to notice that the November days finished up on the chilly side. When we awoke next morning it was to find the canal frozen!

We had experienced moving through ice on more than one occasion so we were glad that we had given ourselves plenty of time to reach Leighton Buzzard, knowing just how much the ice would slow us, even when it was quite thin as now.

Bix crunched her way through and by coffee time we had reached King's Lock.

'We'll go through and moor up for coffee,' Owen shouted from the towpath, as he ran up towards the gates.

I shivered my gratitude, for steering in icy conditions is not only cold but tiring. The tiller has to be pushed or pulled quickly to avoid heading into an extra thick patch of ice and the throttle has to be used more if an extra boost is needed to break up the ice ahead.

When he leaves the boat to work the locks Owen can move about more. If there are long intervals between locks we take turns to steer which means that one of us can warm up down below. However, if we are near a flight of locks, then you would probably see me at the tiller wrapped in anorak, overtrousers, knitted balaclava, two pairs of gloves and long, thick woolly, wellie socks. I'd be stamping my feet and swinging my arms about while waiting for the lock to open. But not Owen! After the second or third session of paddle winding, gate pushing, heaving and tugging, he will put down his windlass, take off his woolly hat, scarf, top jacket and sometimes even his gloves! There was one famous occasion on a cold January day he sweated his way to the top of the Hatton Flight in shirtsleeves while I stood, raw and raddled, on deck.

He watched me steer through the icy surface and after closing the gates walked to the top gates whereupon I popped the kettle on a low gas so there would be no waiting for that much-needed coffee once we were through this lock.

Back on deck I found Owen returning to *Bix* wearing a puzzled expression.

'There's no water in the next pound,' he told me, obviously rather concerned.

'Iced up all the way?' I asked despondently.

'No, it's empty,' was the even more surprising answer.

He set off to explore while I thankfully vanished below to prepare the coffee. After 20 minutes I became worried and decided to lock up the boat and search for him.

The pound connecting us to the next lock is about a mile long but now it looked like an uninspiring, muddy ditch. The reason for its dewatering was explained by a pipe that was being laid across the canal to a new housing estate.

When we enquired how long we would have to wait before the work was completed, we were told only until Thursday. Three whole days! I was furious. We carried the BWB official stoppage list of essential maintenance and this work was not even mentioned. Owen went into Aylestone to telephone the BWB and was told that the work had been started on a spur-of-the-moment decision by someone on the section. The contractors had planned the operation to last two days, but it had taken almost three weeks. We were told, however, that the canal would definitely be opened by Thursday morning. The BWB official added that they did not think anyone would be boating at this time of the year. I could only wonder why they issued annual licences!

There was nothing to do but contain our impatience as best we could for the canal here is less than attractive. Leicester itself possesses a pleasant mooring near St Margaret's Church but, like most cities and towns, its fringes of wasteland, where we were now stranded, attract vandals and loads of rubbish. In fact, on the previous occasion we had boated this section we had tried unsuccessfully to stop two young boys setting fire to the empty house at the very lock where we were now stuck.

For the next three days we made sure that the boat was never left unguarded. As the temperature dropped, so did my spirits. On the Wednesday afternoon I had almost completed the ironing when a familiar sound made us both stop what we were doing; a boat was travelling towards us! Outside we saw that the pound had not only been filled but had had time to freeze over through which two boats were now pushing their way from the other end. We were overjoyed to recognise our old friend Mike Grant who was taking two of his hire fleet to their winter quarters at Brownhills for their annual maintenance. Unknown to either of us we had been waiting at opposite ends of the pound. Mike was even angrier than we were at the unofficial stoppage for he was on a business trip and had allowed only the normal travelling costs for his crew. Thanks to his five-day wait he was having to make alternative arrangements to get the boats along the last leg of their journey.

After a quick cup of coffee together we set off in opposite directions: we to use the channel that Mike's boats had broken for us, and Mike to reach the River Soar and running water as quickly as possible.

We got as far as the second lock, Gee's Lock, before we became well and truly ice-bound. I had to go up to the bows and break the ice with the boating pole while Owen reversed out. He then put the throttle full ahead but we moved only two feet before shuddering to a halt. Thus, painstakingly, we travelled to Glen Parva where we stayed the night.

Next day we managed only 100 yards in four hours whereupon, with shoulders aching from ice-breaking, I gave up. We spent the rest of the day warm and cosy inside *Bix* but far from relaxed as we wondered if we could possibly now arrive at Leighton Buzzard in time. The following three days were spent battling through to Foxton with me ice-breaking in the bows and Owen in the stern throttling backwards and forwards to move us a few feet at a time. The temperature rose slightly after that and we made fairly good progress back to the main section of the Grand Union at Buckby, but as freezing fog descended most afternoons we had to stop travelling at about three o'clock each day.

The remaining 45 miles to Leighton Buzzard took us five days and once again it was too cold to steer for more than an hour at a time. We arrived with four days left to shop, cook, and generally prepare for Christmas on *Bix.* It had been a frustrating, irritating time — but at least we had made it.

11

Our family Christmas on board *Bix* was magic, although the presence of the family brought to the surface mixed and nostalgic feelings about our present life. As Owen and I struggled to fill Erica and Dylan's stockings without waking them I thought of previous Christmas Eves we had enjoyed at the farm.

'If you hadn't made such a thing about it, Rebecca would not be insisting that her children have the Bryce traditional Christmas,' Owen whispered with a wry grin, making a quick dive for a damned tangerine that kept rolling off our duvet! We were filling their stockings in our bedroom, while the grandchildren wrapped in sleeping bags dreamed the night away on the lounge floor, their mother on the bed settee alongside.

'It's impossible for me to do the stockings — I'll wake them up,' she had pointed out earlier, so that was how we came to be whispering and giggling like two kids ourselves as we tip-toed among pillows, cushions and sleeping bags trying to sort out which pair of arms or head we could see through the gloom belonged to Erica and which to Dylan. Hopefully, we placed the stockings at the foot of the sleeping bags and returned sleepily to our own bed.

I swear I had only just shut my eyes when an excited shriek of 'Happy Christmas,' shattered my sleep and with a thump Dylan landed on top of us. It was Christmas morning!

Even the problem of the Christmas tree was solved. Our daughter, Linda, had bought a tree so large that Ian could not stand it upright in their lounge, so on a pre-Christmas visit to us they had brought with them the top two feet which now proudly decorated our coffee table.

The weather apologised on Boxing Day as we walked in sunshine along the towpath towards Linslade. Winter brightness picked out the boxes of multi-coloured, wriggling bait that surrounded the fishermen on the banks. So much tackle and glistening new cane baskets lined the waterside, that I think every male in Leighton Buzzard must have received fishing gear as a present.

All too soon the family had to depart leaving the boat strangely quiet after the gaiety contributed by our grandchildren.

Owen and I had decided to look for work in Leighton Buzzard as soon as the holidays were over. We started with the local paper, and made enquiries at local farms and boatyards. Within a week I was working in a baker's shop and Owen had found work with a freight company handling car engines. We were prepared to spend the next two or three months working weekdays and moving *Bix* at the weekends. Sometimes we would bring the boat back to the same mooring or, alternatively, leave her in another part of the canal and cycle to work. That was the plan!

Throughout nearly eight weeks of January and February, however, we found ourselves iced up by the bridge in Leighton Buzzard. With four inches of ice on the canal, it was impossible to move the boat. Luckily, we were near a water tap and sanitary station, but gas supplies were a problem, first because of a lorry drivers' strike and secondly because the boatyard supplying our gas lay two miles further up canal.

We were more than grateful to the owner of the boatyard, Mike Grant, who delivered new cylinders of gas to the boat regularly throughout the winter. Not once did he let us down, even when the roads were snowbound or treacherous with ice. He always got through, a friend indeed.

One morning we awoke to find it impossible to open the doors at either end of the boat; the wells were full of snow. Through the windows it was difficult to see where canal stopped and towpath started: a pathetic, hunched-up bundle of grey feathers, a heron, peered into the bright white counterpane hiding his breakfast. With a few hefty pushes we managed to open the back door and soon shovelled a path clear so that we could get on and off the boat.

During the hard weather our forlorn heron came each morning to stand by the side of an alder tree, desperately searching for food. I was sure he was starving and wondered whether he would eat shop-bought fish, as he must have been used to swallowing his meal as he caught it. There was only one way to find out, so I bought a couple of whiting, cut them in half and threw a piece towards him across the ice. His quick eye followed its progress but he made no move towards it. I waited, then threw a second piece which landed nearer. He stretched out his neck and as it grew longer and thinner I thought it must break off, it was *so* painfully thin. But suddenly his beak snapped with a click and the fish had gone. The bird looked around anxiously, perhaps wary of a trap, then with a very undignified slithery, sliding walk he reached the other piece of fish.

That was how we came to have a heron to breakfast for a few weeks, although we failed to tame him; in fact he disliked eating if we watched him from outside the boat, so I merely threw the fish across the ice and supervised his meals from a warmer spot — my galley window. He thanked us in one way though by roosting on the boat a couple of times. What better security guard could we ask for?

Most evenings we read or arranged our week-end lecture courses, booked as we were twelve months ahead for sessions on our subjects of jazz and canals. One evening I emerged from the depths of one of my favourite Richard Jefferies stories to hear a murmur of voices growing louder as they approached *Bix*. I sat back in the warm, cosy cabin and wondered who on earth would be walking the towpath on such a cold, dark night.

Then I realised that the commotion was coming not from the towpath but from the canal! Pulling aside the curtains revealed a group of people with torches walking along the frozen canal; en route from a local housing estate to town. The ice was thick enough for others to emulate them during the following week, and as the children were still on school holidays we had plenty of activity all around us.

Even had I felt the urge to skate or even just walk on the ice I would have been unable to do so. Navigating *Bix* through the ice on our journey from Leicester had damaged my left shoulder and arm. A visit to an osteopath revealed a frozen shoulder! It was an apt enough description but one that would attract a smile rather than sympathy. I tried massage, manipulation and even acupuncture, but was told that it would be 18 months or so before I had full use of my arm again — and if I tried to use it before that period I could damage it permanently. So the weather was a blessing after all; no steering! It was just as well for it was the left arm I use for all throttle work which often calls for quick forward or backward thrusts when negotiating locks.

For the first time I found our life disadvantageous. Unable to have regular treatment for my complaint during the next year or so I experienced much more distress than if leading a normal existence on dry land. However, life still had much to offer and both of us looked forward to moving on as soon as the weather permitted.

In January our son Jan had bought his first house which we had arranged to visit. We left the boat for a week-end and hired a car, picking up my 85-year-old mother who also insisted on seeing her grandson's first home at Maidstone. When we arrived at her flat, she passed on a message from our daughter Marny. Could we call in to see her before going to Jan's as she had some curtains for him. Tunbridge Wells was a bit out of the way but as we had not seen Marny or her husband and the two grandchildren for some time we agreed to divert. When we arrived at Marny's I spied through the front room windows Rebecca sitting with Erica and Dylan. They also lived in Tunbridge Wells and had probably popped around to catch up on our news. Before we reached the front door it was opened by four-year-old Benjamin who with the help of his brother Daniel dragged me into the hall shouting, 'Happy birthday, Iris!'

It was a surprise party for me! As one of a family of secret party-givers I really should not have been so completely overcome, but as I saw Linda, Ian, Marny, Keith, Rebecca and Jan (had he really bought a house?) plus the six grandchildren, I could not prevent the happy tears flowing. I looked at a smiling Owen, and an even broader-smiling old lady, for even my mother was in on the secret. What a lovely day we had even if I did not see Jan's house until three weeks later.

We gave our first 1979 week-end lecture in February when our friend John Palfreman with his daughter Ann took over as stand-in crew of *Bix*. John, a former neighbour of ours in Kent, was keen on canalling and had taken this chance of experiencing boating in winter conditions. Luckily a thaw had set in and as we waved goodbye to them on Friday lunch-time we could see that the ice was now wafer thin.

On the Sunday afternoon our route back by car to the boat followed the canal past Bletchley and Soulbury. The locks were frozen again. The next stretch of canal was also solid. Then we spied our boat snug on her old mooring. A warm cosy cabin awaited us with kettle aboil. John related his saga; he had travelled one mile in two days — and half of that in reverse!

Half a mile from our mooring was Leighton Lock and *Bix* had travelled easily through the ice to enter it. John's trouble came when he tried to leave the lock, for the ice at the lower end was so thick that after half-an-hour's effort he realised he would never get through. As it was impossible to turn round the boat, he had to reverse out of the lock and return through the channel he had cut on his outward journey — in the same gear! Time taken was four hours — and the rest of the week-end Ann and he had spent walking not boating, but our log still read,

'It was super, but hope to do more boating next winter.'

Has he set a record for the least distance travelled on a week-end trip?

At the end of March we bade farewell to Leighton Buzzard and the friends we had made there. It was time to take to the open seas and, yes, I do mean sea!

12

It was at our farewell party, in 1974, that Willy had first mentioned the idea of putting to sea in *Bix*! Cyril, the landlord of the Black Horse, a pub near our farm in Kent, had laid on a marvellous spread, home-baked bread, home-made cheeses and home-killed meat, the way we liked to do things in our neck of the wood. The pub had been a regular rendezvous on Sunday mornings for many folk when Owen's band played in the bar.

There was no public transport to that pub. The whole village, a scattering of houses and a couple of farms apart, consisted of a church, a pub, a small school and an even smaller post office. But the crowds came to hear jazz every Sunday; even when the roads were blocked with snow we still had an audience, and not all youngsters either. We played to an age range of 18-80, and we have captured many happy memories of those mornings on tape recordings and cine film.

Some of the regulars in the audience became friends, others just nodding acquaintances. Willy was one of the latter. He was a small wiry man who was obviously interested in jazz, but only turned up about once a month.

'What's this boat of yours like then?' he had asked and, for the umpteenth time we explained.

'And you're only going to do canals?' he queried.

No, we said, and went on to describe the rivers we could ply such as the Severn, Wey and the Thames.

'Well, if you ever want to come on the River Medway, let me know, I'll bring you down river,' he said.

Owen was immediately interested.

'Surely it's a bit dodgy for a narrow boat in tidal waters right out to the estuary, isn't it? She's not built for suchlike.'

'If the weather is right it should be OK. Anyway I'll look after you. I'm a Thames pilot, so don't forget.'

We had forgotten neither Willy, nor his promise, so now four years later we were on our way to London to take him up on his offer.

The trip from Leighton Buzzard to Limehouse Basin was interesting in itself as we had two French visitors with us. An advertisement in *Waterways World* had brought us together; 'French canal enthusiasts offer holiday in their Paris flat, exchange holiday on English canal.'

Here they were, Roger and Lucette, all set to boat through London. They joined us at Leighton Buzzard for a two-weeks' cruise when we enjoyed showing off our knowledge. As we passed the BWB workshops at Bulbourne, we pointed out the 19th-century architecture still looking as good as new, and the adjacent railway cutting built by Robert Stephenson in the 1830s. Then it was the ruins of Berkhamstead Castle that made their cameras start clicking, especially when I told Lucette that it was here, in 1066, that William the Conqueror was offered the English Crown.

Hemel Hempstead was followed by the lovely Cassiobury Park in Watford and, before we knew it, we were talking about the restaurant at Black Jack Locks, once a canal pub and featured as such in the film, *The Bargee*, made in the 1960s starring Harry Corbett and Ronnie Barker, a film, incidentally, that we show regularly at our canal week-end courses and one which reveals how little the canal has changed.

And so through Uxbridge into London proper. When we emerged from the tunnel at Islington and I informed our visitors that we were only five minutes' walk from the Angel tube station, it was decided to stop for a day's sight-seeing. We then took the boat down the Regent's Canal as far as Bethnal Green so as to give them a quick look round the East End of London. Soon it was time to return to Little Venice where the nearby Warwick Avenue underground station was convenient for them to start their journey home.

The time had come to get *Bix* ship-shape before braving deeper waters.

As the news of our journey to the Medway filtered through the family to friends, we received offers of help with the steering. Willy, whom we telephoned regularly for help and advice, finally set Sunday, June 3 as the day of our departure — but only if the weather forecast was favourable. We wanted no wind. A flat-bottomed craft like *Bix* could easily capsize, especially if the tide were running fast.

We planned to leave Limehouse at 6.45 so Ian, our son-in-law who was to be one of the crew, arrived the previous evening to help check the engine, diesel supply, and our mooring ropes, remembering our earlier horrific experience suspended in that same basin when the abnormally high tide of the Thames had broken over the top of the lock gates. We also ensured that both cameras, one still and one cine, were loaded. This surely would be a wonderful opportunity to film my beloved Greenwich from my own boat, and the building on the towpath near Blackwall Tunnel where as a youngster I had been a junior filing clerk. After that I would pass the Charlton barge-yard where my dad had worked for 30 years or more. A mile further on, and Owen and I would look out for the tower of St Mary's Church, Woolwich, alongside the ferry, where we were married.

Owen, Ian and I were up by 5.30 am as excited as kids while we dug into a hearty breakfast, for Willy had warned us we would not stop until we reached Rochester that afternoon.

An hour later Willy's wife deposited her husband, son and our friend John at the gates of Limehouse Basin. I marvelled at the good nature of the woman getting up so early to bring them from their home in Meopham, near Gravesend. Willy gave her a quick wave and almost before the car had moved off he was stepping purposefully towards us. As he crossed the lock gates, a large cargo vessel entered from the Thames, its pilot alongside the captain on the bridge. He waved.

'Hey Willy, didn't expect to see you today, thought you were off shift. You got a job?' the pilot called.

There was a nod from Willy, and his colleague looked around.

'Which one?' he asked.

Willy pointed to *Bix*, tied up under the bows of a large Greek luxury yacht. Laughter and cat-calls rose from the men on the cargo ship's bridge and Willy joined in as he came aboard.

'And I'm not getting paid for it either!' he yelled after them.

As we left the lock and came out on to the Thames, we ran straight into mist, much to my disappointment but obviously to Willy's delight. It was not very thick, but enough to prevent photography.

'Perfect weather for us,' Willy smiled. 'Not a breath of wind; we'll have a lovely, sunny calm day by the time we reach Garrison Point.'

And so we passed Greenwich, Charlton and the Woolwich Barrier, which looked like a grey ghost ship itself, seeing only dusky shadows of buildings suddenly looming into view and then silently slipping away.

My dining table was covered with charts which were constantly checked by Willy, with Owen or Ian by his side. It was reassuring to be in the hands of an expert and as he talked of tides, marker buoys and sand channels we realised how lucky we were to take *Bix* on such an unusual journey.

According to Willy's reckoning we should have reached Garrison Point at Sheerness by 1.30 pm enabling us to catch the Medway tide. As we progressed down river and drew nearer the estuary, the mist lifted. Across the wastelands of riverside Essex we knew that Southend Pier would be pointing far out to sea. Would we be able to see it, we wondered? From Gravesend on it was already becoming difficult to make out landmarks, for we were now in such a vast expanse of water that to us on little *Bix* it seemed that we were already at sea. The tide was taking us much faster than we had experienced before, and yet it seemed an age before the scenery changed; at times we seemed stationary and it was only when other vessels crossed the horizon ahead did we realise that we were indeed moving quite quickly.

As we started our turn into the Medway, we checked watches.

'Not bad,' nodded Willy. 'I told my wife to pick me up at Rochester at four o'clock.'

He went below for a well-earned cup of tea, not made by me however, for professionals like Willy bring their own food and drink with them.

Then two Thames barges hove in sight and I rushed for my camera. As I focused on them and snapped away I spotted figures on the barges leaning over the side. We laughed and waved to their passengers photographing us photographing them.

The tidal part of the Medway was congested with small racing dinghies through Faversham, Gillingham and beyond Chatham. As we headed for the jetty near Rochester Castle the clock chimed four and there was Willy's wife, waiting by the car. With his bits and pieces already packed Willy and son leapt ashore; no fussy goodbyes, just a wave and a grin to acknowledge our thanks. We hadn't even tied up!

Soon we had to pass through Snodland, a village dominated

by its cement works which deposit a covering of fine grey dust on every tree and hedgerow. We knew this place of old, and the neighbouring Medway towns and villages, for hadn't we lived for 20 years or so within five miles of the river? But Snodland viewed from the water proferred a different aspect and even that poor cement-choked district revealed an interesting landscape with an old church tower stark against a backdrop of the North Downs.

We carried on towards Aylesford with its complex of paper mills lining one side of the river. Here there is one place of beauty worth any diversion. It is the Carmelite Friary, the first to be built in England and still a haven of peace and beauty despite its industrial surround. The gold-grey stone walls, reflected in the river below, hid a rose garden that I had trod many times and, I hoped, it would not be long before I was looking around this scent-filled spot again.

The Friary stands near an old stone bridge, a traffic nightmare, so narrow that it can be used only in one direction at a time. On many occasions in the past I cursed Aylesford Bridge when I was delayed to and from our farm, but now I could smile serenely as we glided lazily under its lovely arches.

Allington is the first lock on the Medway and as it is tidal it is open only three hours before and two hours after high water. When we reached it, I was reminded of Boulters Lock on the Thames at Maidenhead where the ever-present onlookers join in the laughter when a newcomer to boating makes a mistake! The Sunday evening we locked through Allington was still warm and sunny and it had certainly brought a bevy of 'gongoozlers' as they are called on the canals, to watch our progress. Perhaps the nearby pub on the towpath had something to do with it as well.

The lock-keeper smiled.

'Crikey, you're a big one, it's nice to see you.'

We were puzzled. He seemed to be expecting us for he continued, 'The people on *My Lady Val* said you could tie up alongside whenever you like.'

'*My Lady Val*? What's that?'

'It's a narrow boat, like yours,' he explained, 'but not so big. Anyway, you can't miss it, they are waiting for you at Yalding.'

The Archbishop's Palace clock in Maidstone struck seven o'clock as we tied up. We were worn out, having travelled twelve hours non-stop. Poor old *Bix*, she had not been put through such an ordeal before, and neither had we!

The next morning brought a surprise visit from our son Jan who called with our post and invited us to dinner at his home in Maidstone. He told us we could tie up almost at the bottom of the street where he lived; well, the one that ran adjacent to it, anyway.

By 10 am I had completed my shopping in what had once been my regular stores, met an old friend and had been invited to visit Coral, my producer at Radio Medway for a reunion broadcast on a programme to which I had regularly contributed. It *was* nice to feel part of somewhere again!

For the first time in four years, it seemed strange to return to *Bix*. I looked at her from the bridge and could not really believe it was not all a dream. I would surely walk across to the car-park and drive back to the farm, a mere 20 minutes away. But no, there was Marny, with Ben and Dan on the stern of *Bix*.

'I've just put the kettle on mum, you're in time for coffee.'

Even the family noticed a difference.

'I can't get used to popping over to see you in the evening,' said Rebecca.

So with Marny.

'It's great to have you so near,' she told me. 'I can shop so much quicker now I can leave the children with you.'

She was a regular caller of course, and I loved every minute of it.

Quite by chance we had arrived at the right time to meet the newest addition to our family — a future daughter-in-law. Jan had already brought Linda to dinner aboad *Bix* at Little Venice when she had been introduced as a girl friend. Now, we were moored a few minutes away from a favourite restaurant in East Peckham which we had used frequently during the years we had lived nearby. Jan and Linda were with us for a week-end trip and he had reserved a table for the Saturday evening. I was surprised to find that he had booked for six. The extra pair turned out to be Linda's mother and father, Don and Joan. Before dinner while Owen and I were getting to know our two new friends, Jan and Linda left us for a few moments; when they returned a glowing and smiling Linda made for her mother to show off a sparkling new engagement ring. The restaurant did us proud that night, and it is more than ever firmly entrenched as one of 'our' places.

We had intended to stay on the Medway for a month but it was to be ten weeks before we managed to tear ourselves away. We may have lived and worked near the river for all those years and boated parts of it in a friend's small dinghy but it was only now that we realised just how beautiful a waterway it was that wandered through the lovely Kentish countryside.

The tidal Medway, which comes under the administration of the Medway Ports Authority, starts between the Isles of Grain and Sheppey, and runs past Faversham, Sittingbourne to the Medway Towns, Chatham Dockyard and up to Allington. The rest of the river is controlled by the Southern Water Authority. Both authorities issue extremely useful booklets for boaters.

The non-tidal Medway is sheer delight; castles, palaces, markets, churches, pubs, strawberry fields, hops, oast houses and the most beautiful old bridges at Teston and East Farleigh.

To my delight five minutes' walk from Maidstone Bridge I found an osteopathic clinic where I had weekly treatment for my damaged shoulder. The relief so obtained from these sessions was much appreciated as I had reached the stage of being unable to put my left arm behind me or above my head. I took double the time to dress in the mornings and rediscovered how often a woman's clothes do up at the back. My method was to fix bras and skirts from the front and then tug them round. Blouses and jumpers which had to be pulled over my head I simply gave up wearing. I could get them on with a one-handed struggle, but it was too painful to remove them again, even with Owen's help.

Boating doesn't help to cure an ailment such as mine though, as I found to my cost during our last week as we locked through Hampstead Lock at Yalding. The lock is the second deepest on the Medway and an inexperienced girl in a small boat alongside us got into difficulties as the water swirled into the lock. The craft started to swing swiftly towards us whereupon I hung our fenders over the side to prevent their fibreglass hull crashing into our three-eighths inch steel sides, and then threw a rope for the girl to attach to their stern cleat so securing them alongside *Bix*. As I completed these manoeuvres a sudden pain at the bottom of my shoulder blade shot up my arm to my neck. I was back to square one again! Although my shoulder was painful at times there were moments when I was able to forget it. I soon learned to relax when I felt the pain coming on but knew that a permanent cure needed specific help. As I had finished my weekly treatment at the clinic, the osteopath gave me a series of exercises for the boat.

The mysterious message we had received from the Allington lock-keeper about *My Lady Val* was solved when we cruised past the boatyard at Yalding where we spotted a 45ft narrow boat, splendidly decorated, bearing her name. A smiling, buxom lady aboard greeted us enthusiastically.

'Super to see you; we read all your articles, know all about you. Come and tie up and have a coffee.'

It was Val Harmer, who with husband Tony and son Steven lived on the boat but who visited other canals on day trips by car! They possessed every book and magazine covering the historical, industrial, wild life and leisure side of those canals, but they had yet to boat on them! Apparently they had bought the hull of their home and had it transported by road to Yalding where Tony had completed it and now worked in the boatyard where the mooring went with the job.

During our stay on the Medway we met several times and enjoyed the ever-amusing company of ever-smiling, ever-chatting Val. It was obvious from the questions we were asked that they were about to follow in our wake.

'We reckon we'll be on the move next year,' said Val. 'Tony has decided to give up his job here and we'll do the same as you, travel around and take temporary work.'

We had heard this sort of talk so many times before, but here, I felt, was a couple who meant it. For one thing, Val had already made enquiries about continuing the education of their 14-year-old son by post.

Our regular mooring in Yalding, however, was not alongside the Harmers, but at an old thatched pub on the bend of the river. The Anchor was the home of an old friend of ours, Robin Brenchley, known to all as Brench. A great jazz fan he gave us a warm welcome to tie up at the bottom of his garden so that as many hours could be spent talking and listening to jazz, as when we lived on the farm.

Almost at the end of the navigable part of the Medway, in Tonbridge, we met a delightful couple living on a narrow boat. A few permanent moorings are beautifully situated near Tonbridge Castle, in the park laid out by the riverside. I have always maintained that I could not live on a residential mooring but after seeing Michael and Dorothea's floating home in such glorious surroundings I reckon I could give Tonbridge a try one day. When I asked if their boat *Amaryllis* was named after the story by Richard Jefferies, I noticed how Michael's eyes lit up. It's not often fans of this Victorian author meet and Michael became even more interested, not to say envious, when he found out that I actually had some of Jefferies' books with me on *Bix*, including *Bevis and Mark* his famous children's story. Michael had only seen copies in libraries before. Now he's looking around jumble sales in earnest, as that is where I found mine!

The weeks sped by and we were glad that on applying for our licence to boat the Medway we had taken out one for a year. It cost only £12 and we really couldn't believe that waterway travel could be so inexpensive. It must surely be the cheapest waterways licence in the country; certainly we have boated nowhere else so economically. We were told that we had to thank William the Conqueror for this as he had given the Medway to the people of Kent as a free river. The licence money is really a lock pass. Incidentally you can buy a daily, weekly, monthly or annual pass. And if you want to boat only between locks there are some pretty long stretches where you can slip your boat in and out free of charge. There are also plenty of public slipways.

We planned to leave the Medway on July 22 but this news was greeted with cries of horror from the Harmers.

'Please stay for the Maidstone carnival. It would be great to have another narrow boat there.'

So we stayed. And what a marvellous day it turned out to be. The town is to be congratulated on using the river as a backdrop to the carnival. There were boats galore to be seen, some moored three abreast, from small fibreglass dinghies with flags flying, to floating cocktail cabinets, and even a glorious schooner, fully rigged, as well as many rowing boats and, of course, a couple of canal boats. A presentation parade of the year's Kentish beauty queens took place by water, each arriving with her attendants to be greeted by the mayor in his boat moored outside the Archbishop's Palace. *My Lady Val* prouldy bore the beauty queen of Sheppey.

It was a fortnight afterwards before we managed to uproot ourselves. Willy had presented us with one of his charts and persuaded Owen that he was more than capable of skippering

Bix back on his own. As both Owen and Ian had squeezed Willy dry with questions on the way out, I had no qualms about venturing forth without a professional pilot.

Our crew for the return journey included Jan and his friend Graham. Ian was to be picked up at Rochester, which ensured four helmsmen as before.

The tidal lock at Allington should have been operating by 9 am on the Sunday we had chosen to depart but in fact it was nearly 10.30 before it was possible to lock through.

Ian was waiting at Rochester with Linda and their two children who had come along for the ride, and soon we were enjoying the slight swell as we rode the sunny waters past the oil refineries on the Isle of Grain. The chart was regularly checked to identify buoys marking the estuary sand-banks. Some buoys housed warning bells which rang at intervals, amusing the children, and pleasing the adults. Soon we could see Southend on the horizon and the children were trying to spot the hill near their home at Laindon where they often play and picnic.

All seemed well until Owen checked our time as we approached Gravesend when he realised that it would be dark before we reached the Woolwich Barrier. There was no hope of getting to Limehouse and the safety of a canal mooring that night. Waiting for the tide at Allington had put back our arrival, and we had forgotten that the nights draw in earlier at the end of August than they do in June when we had arrived.

Our Thames guide indicated a mooring at Erith causeway was possible at all times so we decided to stay there. However, when we reached the long jetty stretching out into the river it was obvious that we could not moor alongside the causeway. A bank of mud lay between us and the causeway so we hitched up to the jetty.

Luckily, we are familiar with Erith and knew that a police station was located nearby. Owen called there to ask them to make three phone calls; to the river police to ask where we could moor, to Woolwich Barrier whom we had notified that we would be coming through; and to Limehouse Basin who were also expecting us to lock through that night. And at this stage our crew abandoned ship to return home by train.

Owen was away for a worrying ten minutes leaving me on board alone with the tide making fast. The jetty we had tied to was now under water and *Bix* was in danger of floating across it.

If I untied, I would be carried away upstream on the current. Even with the engine to help I would have had great difficulty in turning round to pick up Owen.

With relief I saw him returning, the water lapping his sandals as he ran along the jetty. He jumped aboard and cast off the ropes indicating that we were to head for a line of empty barges midstream. The river police had urged all speed adding that the only alternative was to continue in the dark and try to moor at Woolwich, or even Greenwich, another two hours at least. We steered for the barges!

With the memory of that wait outside Limehouse Basin tied to barges two years earlier I wondered how on earth I would be able to attach the mooring rope to a cleat, as Keith, a six-footer, had been on board to help me on that occasion. When the barges loomed up I threw my rope and watched the loop drop safely over the stern cleat! I bowed gracefully to my non-existent gallery.

Darkness fell swiftly, more like the equator than the outskirts of south-east London. Navigation lights twinkled on a ship moored 100 yards down river; we were glad that our well-charged batteries enabled us to keep on a light in every room, a luxury that is never needed when cruising the canals. We also hung a hurricane lamp from the tiller and attached a fluorescent torch to the bows of *Bix*.

By the time we had finished our evening meal it was ten o'clock and bed beckoned for it is certainly hard work steering a narrow boat on a river. But there was to be little sleep for us that night.

An hour later a sudden crash against the side of *Bix* tumbled china and glasses to the deck while the pictures swung crazily. Then the boat took another bang. We leapt from bed and with difficulty walked through our rocking lounge.

From the boat's stern we saw the cause of the commotion: the tankers and cargo ships which had been waiting in London for the tide were now coming down river. As they steamed past their wash hit the empty steel barges which in turn cannoned into us. The noise was deafening but it was one we had heard before and had forgotten. The previous drama had been played in daylight when we had whiled away the hours waiting for Limehouse Lock to open by playing Scrabble and reading to our grandson. Now, too late, we recalled the cacophony as lasting an eternity. So we made coffee, read, promenaded the deck to watch the shipping pass, read more and, finally about 4.30 am in a lull, we returned to bed.

We awoke to a silent world: an early morning sun turned the river surface into a pale, silvery, rose-coloured stage. The Thames was as still as a mill-pond, uniquely beautiful. No gulls screamed overhead, no hooters, engines, train whistles in the distance. Just the two of us with the world to ourselves. We quietly untied and set off on the last leg of our river journey. It was 6 am. Although lacking sleep we felt revigorated as we travelled through what is normally a bustling, commercial river, entirely alone. A couple of swans swam out to greet us as we passed the huge Ford display sign outside their Dagenham factory, our sole companions until we rounded the bend of Galleons Reach and saw the Woolwich ferries crossing ahead.

A launch chugged towards us from the south side of the river; it was a pilot boat directing traffic through the Woolwich Barrier which is London's new flood protection scheme, nearer Charlton than Woolwich, but nonetheless called the Woolwich Barrier. Sadly, the pub used by my Uncle Bill throughout his life as a lighterman which he knew as *The Lads of the Village* is now called *The Barrier.* How long will it be, I wonder, before people forget that Charlton was once a village?'

The pilot requested our destination.

Although we had planned to return via Limehouse, Owen suddenly replied, 'Bow Creek.'

'Thought it would make a change,' he mumbled in my direction. I stared at him wonderingly.

'We've been through Limehouse before but we've yet to see Bow Creek,' he explained.

We had yet to find it! Thank goodness we had Willy's chart showing Bow Creek, unlike *Nicholson's Guide to the Thames* which indicated various streams and inlets entering the Thames but not one called Bow Creek. Even with a marked chart it was not easy to spot which narrow gap of water led to the creek and we could easily have found ourselves sailing into the Royal Victoria or even Albert Dock by mistake.

A launch for dealing with downstream Barrier traffic was moored at the entrance to a small inlet. As we approached we saw the words 'Bow Creek' in large but faded lettering on the side of a building. The pilot hailed us with astonishment written on his face. His outspoken comment was,

'I've never seen the likes of you this far down before. Where have you been, and where the hell are you going?'

We tied up alongside, relating our story while we breakfasted with him. He telephoned Bow Locks to see if there would be sufficient water for us.

'With your speed,' he told us, 'the tide'll overtake you before you reach the locks, so you're OK.'

'Wish my old woman was like you,' were his parting words, 'but she gets sea-sick looking at a bloody bath of water.'

As we started on these waters new I thought that mud would be a better word than water for them. We had enough water under us as we only draw 1ft 9ins and we made fairly easy progress although bordered by some five feet of mud on either side. There were several hairpin bends and we were surprised as we swung round one to behold two large ships aground by the side of a wharf. Obviously at high tide cargo boats could come this far.

Huge colonies of swans preened themselves on the sunny mudbanks, and we noted coots swimming fussily in and out of the weeds growing on piles of driftwood. A lump of polystyrene drifted against us, on it three fluffy moorhen chicks cheeping non-stop for mum who swam by their side while she foraged for titbits from heaven knows where. And all this a few yards below the surface of the road from Blackwall Tunnel to Hackney!

Once through Bow Locks we were in familiar waters leading to the River Lee.

Would we recommend a trip to the Medway for canal boaters?

The answer for first-timers is 'yes' if you employ a pilot, weather forecasts are in your favour and everyone wears a life-jacket. They may not be normal wear for most adult canal boaters but are essential on tidal waters.

Would we do it again? Most certainly 'yes' in Owen's case. But me? Well, I must admit that there were times during that twelve hours when I was bored. I could not steer, as it was too tiring, and I could not settle to read, but the end product, the River Medway, was so rewarding that I would probably answer affirmatively too. In any case, there would be such a rush of volunteers to help Owen that I would not be missed if I went by train and caught up with them in Maidstone!

13

There are many names for it — but I call it Sod's Law. It means, for example, that when we decide to take *Bix* to one place, that we are immediately booked for lectures or invited to friends in the part of the country we have just left! The morning after our arrival in Maidstone, Owen was offered a band job for 26 Wednesday nights in Mablethorpe, on the east coast not far from Cleethorpes. We needed the excellent money so he took the booking and although the fee included travelling expenses, how much easier it would have been if we could have moved the boat nearer the job. And now that we were back on the Grand Union, heading once more for Leighton Buzzard, our son's wedding was announced for September. As this was to take place in Otford, next door to our old home Wrotham, we found ourselves back in Kent a few weeks later. That journey, however, had to be made by car.

After the wedding we started to think of our winter cruising plans. Much of the canal system is closed for weeks at a time during the winter months for the repair of lock gates and the re-lining of lock walls. In some cases where tunnels have become dangerous canals are shut for a year or more.

We consulted the BWB official stoppage list and found that the River Stort had the least work taking place that winter. We had already spent an enjoyable three months or so cruising the river from November to February a couple of years before when I had fallen in love with Bishop's Stortford so we decided to make our way there.

Owen had previously worked as a labourer in a local steel works and Andy, the foreman, had offered him work if we ever returned. A phone call confirmed the job was still available, so we pushed on as quickly as we could.

I also had a plan for my shoulder. If we were to be near a hospital for two or three months, I could continue with treatment. A doctor friend of ours who had noticed how little I now used my left arm, remarked, 'If you didn't lead such a peripatetic life you could get relief from regular physiotherapy.'

We had an easy trip up the Stort but at Twyford Mill Lock we were warned of delays. At South Mill Lock a couple of earth-moving machines and a dredger were building flood channels which should end the inundations suffered every year in Bishop's Stortford, Harlow and adjoining villages. We had hoped to moor near South Mill but were advised to proceed into the town itself, at the end of the navigation channel. We were warned that two barges would be operating constantly between the bridge and Sainsbury's car park and South Mill Lock, carrying away the mud dredged from this stretch. Additionally the towpath was to be closed temporarily, making it impossible to get into town.

We secured alongside what had once been a wharf behind Pickfords Travel Centre. Years ago it had been used by their own boats when the Stort was a working river. Little were we to know that several weeks would pass before we could leave that mooring; in fact during the next three months we managed to cruise only as far as Hallingbury Mill and that but once or twice. Luckily a garage backed on to the river nearby whence we obtained our weekly water supply from a handily placed tap.

The work was a joint effort between the British Waterways Board and the Thames Water Authority and not on our stoppage list. When I complained that the new banks of the river built of stones embedded in concrete offered no chance of wild flowers growing there or animals making a home by the river, the TWA said that that part of the work was planned by the BWB. They in turn, of course, told me that they had had no say in the matter whatsoever!

Owen started work next morning while I made my way to the Job Centre to look at the few vacancy cards. I was not hopeful. The only temporary job available was as a lunchtime part-timer in a pub, cooking and serving light meals and washing up.

The interviewer told me the job would not suit me as it was at *The Bakers Dozen*, a cellar bar at the top of the town which appealed mostly to the youngsters. With visions of psychadelic lights and noisy electronic music, I agreed. I walked fruitlessly through the precinct to see if any of the shops were taking on extra staff for Christmas and on my way back to *Bix* passed *The Bakers Dozen* from which emerged a cleaner whom I asked if I could have a look inside. She referred me to the manager downstairs who was busy in an attractive bar, cleverly adapted from a cellar and small tunnel that ran beneath a courtyard, and had once connected the hostelry with a brewery some two hundred years earlier.

Nick, the manager, in his early twenties, was pleasant and did not seem to mind my asking questions about the job. I looked around for the juke boxes and multi-coloured spotlights. There was none.

The work sounded OK to me and presumably Nick approved of me, for I started straight away on what was to be a three-month stint. Peggy, in charge of the bar, was a lovely person and the customers, mostly local businessmen, were far removed from what I had been told by the Job Centre. We always had a crowd of teenagers on Saturdays but none was troublesome in any way. If it were noisy there then it was usually chatter and laughter, especially during Christmas week.

I enjoyed my work especially as the hours allowed time for twice-weekly physiotherapy for my arm and shoulder at the local hospital. I had no difficulty in finding a doctor ready to take me on. The morning I met Dr Kent he examined me and said he would write immediately to the local hospital for an appointment and I would hear directly from them. As I dressed he wrote his letter and then asked my address. Now, I thought, trouble starts! I explained that my mail would have to be sent c/o the local Post Office and went on to describe our nomadic existence.

'How long will you stay here then?' he added.

Knowing that he must be thinking that it would not be worthwhile to start any treatment I told him we should not leave before the end of February.

'Ah, that's good,' he smiled. 'Could you talk at our Rotary Lunch in January then?'

And I did and was made such a fuss of.

As their meetings took place in the house where Cecil Rhodes was born I had the added bonus of being taken around and shown this interesting family house.

The hospital treatment worked splendidly and later in the year, just as my physiotherapist had promised, I found that I could use my arm normally once more.

We said goodbye to our Stortford friends at the end of March and not February as we had planned because of delays in completing the flood channel dredging. We were in no hurry: the rest of 1980 lay ahead of us.

14

By Easter we had reached the River Wey, entered from the Thames below Shepperton Lock. At first it is not easy to distinguish which stretch of water leads to the Wey as on quitting the Thames four different channels face you. The navigable one, however, is indicated by a sign and within a few yards of this a 'stop' gate can be seen which enables boats to enter the first lock should the level of the Thames be too low, or their own draught too great for them to pass over the cill. If the gate is open all boaters must tie up by the steps and report to the lock-keeper of the Wey's first lock, Thames Lock. If there is insufficient water for your boat, the lock-keeper will close the 'stop' gate and draw water from the top gate paddles of Thames Lock.

We found the gate open and had plenty of water to enable us to reach Thames Lock without water having to be drawn off. The river, which is navigable from Weybridge to Godalming, is owned by the National Trust. One or two of the 14 locks are difficult to negotiate due to fast-flowing weir streams, particularly at Papercourt Lock where the boat must be handled carefully to avoid the powerful current which can sweep you against the lock walls.

On this rural route supplies are best picked up in Weybridge and Guildford. There are a few small shops near New Haw Lock but as this is only a couple of miles from Weybridge we had no need to use them. It is certainly worth stopping there if the owner of the lock house is at home. He is a fund of knowledge of the history of the river and I shan't forget admiring his rose-covered cottage while he told us that the kitchen was haunted — by a tea-cloth of all things! He had seen it float across the room several times; one morning he had found his wife on the floor in a faint after the spectral material has whisked past her face. . . .

Guildford moorings are splendidly situated and before Owen had entered our time of arrival in the log-book I was across the bridge to the Yvonne Arnaud Theatre to enquire successfully about tickets for that evening's performance.

A couple of months before cruising the Wey we had received a letter from a Mrs Scott who was a member of the Inland Waterways Association and who, along with her husband, worked actively on the restoration of the old Basingstoke Canal. The junction of this, at present, unnavigable canal is just past New Haw Lock. She had written to explain that after reading some of our articles in the waterways press she realised that she had been known to us in 1949 as our regular baby-sitter! She now lived in Guildford and invited us to contact her any time we were nearby.

Like so many wonderful people who work hard restoring our waterways with little reward, Jean and her husband Gordon did not possess a boat of their own, so we managed to fit in a short trip of their home river, the Wey, whilst we were with them. Altogether we spent six pleasant weeks cruising between Weybridge and Godalming, three days of which were spent near the village of Send when we found the flood-gates at Worsfold closed due to a swollen river. As we cruised along we had spied the square tower of Send church from two or three different vantage points because it lies in a wide loop of the river. Now we had time to explore not only the church but the village. Alas, church apart, it failed to excite us in any way.

By the side of Worsfold Gates stand the workshops of the Wey Navigation Authority where we had some informative chats with the workmen who talked of the days when the river was busy with boats carrying timber to London, returning with corn for the riverside mills. Just above Weybridge, Cox's Mill is the sole survivor still using the river as a means of transport. The surrounding countryside, despite a few stretches of urban development, is pretty and most of our journeys were against a background of copses, farms and even a couple of historic ruins, Newark Priory at Newark Lock and, at St Catherine's Lock, a chapel dedicated to the same saint stands atop a nearby hill. If you want to see the chapel at close quarters you can reach it by climbing a steep, red sandy cliff, a few minutes' walk from the Guildford Boathouse. If that sounds too energetic there is a footpath near the main road.

St Catherine's Lock is in open countryside; all the other locks seem to be hidden by sharp bends with attractive but awkwardly placed trees hiding not only the entrances but also the noticeboards directing you away from the weirs! The scenery between Triggs and Bowers Locks is inviting for you are looking at Sutton Park, home of the late Paul Getty but, alas, there is no public right-of-way from the river.

One or two of the locksides are turfed, something of a surprise to us as all other locks we use have brick or concrete sides. The sloping turf decreases the size of the lock and I was glad that at no time did we ever share a lock with another boat, especially when we were working down river as I like to keep *Bix* bang in the middle of the lock to avoid getting caught on the jutting edges. As interesting as it is to see these original locks I prefer the brick-sided variety; it is too difficult disembarking when the sides are sloping wet turf!

Broadford Bridge has only six feet headroom, so care must be taken. Just above it is the entrance to the Wey and Arun Canal which runs to the River Arun at Pallington, near Pulborough. Although currently unnavigable volunteers are hopeful that their efforts to clear it will soon succeed. At a cost of £20 for an annual licence, the River Wey is good value and we shall certainly grace it again with our company!

Later in the year, we cruised the 10 navigable miles of the Kennet and Avon Canal, that is from Reading to just past Tyle Mill Lock. There are other sections now open for boating but to date it is impossible to cruise the entire length of this waterway which was once the main east-west canal route linking the Thames to Bristol. A vigorous canal society has restored much of it, and if the remainder is as beautiful and rural (apart from Reading itself) as the length we travelled, then a great deal of pleasurable cruising is in store for future boaters.

We entered the canal from Kennet mouth on the Thames and immediately were swallowed by Reading's new building developments and industry, on both sides. Enormous gas-holders and a busy railway line contrast dramatically with the pretty house and garden at Blake's Lock, the first to be reached.

Going through Reading we found ourselves twisting and turning and at one sharp bend we had precious little room to round without scraping our sides. The canal is so narrow in places that I wondered how the local 70ft boat, used for day trips and outings, manages to navigate this part. A notice warned us that one particular stretch was controlled by traffic lights but on reaching them they were out of order. As we were leaving this awkward length of waterway we saw another notice informing us that the lights were not working! Luckily we met nothing coming towards us.

Burghfield Lock is turf-sided, another surprise as the four previous locks were of brick. We moored for the night just below the lock and decided to visit Knights Farm Restaurant, half a mile south of the bridge, so our guide book said. Because of alterations due to the construction of a motorway we walked a mile before realising we were on the wrong road. Eventually we found the restaurant, now almost hidden at the dead-end of what was once the original little country road.

Our efforts were splendidly rewarded. At the restaurant a pretty young lady acted as hostess — yes that's the word I must use as we really did feel we were being entertained at a friend's house for we were the only customers that night. And so we were greeted and made welcome, shown upstairs to a cosy bar, informed by our hostess that her husband did all the cooking and would we prefer to listen to ballet music by Tchaikovsky or Beethoven's Pastoral Symphony while we ate. She had bought cassettes of them that afternoon and could not wait to hear them herself.

We said we would start with Beethoven and leave Tchaikovsky for the brandy. And a splendid meal it was. Even the weather was on our side and afterwards we walked home through a warm moonlit June night. The next morning we were awakened earlier than we would have wished by two Canada geese who had brought along their combined families of 23 goslings to have breakfast by the side of *Bix*. . . .

Unhappily for boaters, the new M4 runs close to the canal around the Burghfield stretch, but we found a fairly peaceful mooring near the bridge.

At Tyle Mill the BWB have landscaped a pleasant mooring site with a small shop and car park for visitors. When we arrived the shop was closed and the only people we met were two old men taking their afternoon stroll. If the place can shut in June, I wonder when it opens for business?

Friday, June 13 proved a lucky day for me. We had moored for lunch near Southcote Lock, and I walked along the towpath to see if I could discover what a tumbledown group of buildings by the lock had once been. A track led down the side of a wall to the back of several brick chambers that had obviously been part of an ancient water-filtering plant. The brickwork had crumbled away and the floor was a mass of rosy Herb Robert. It seemed to be a new strain of the little wild flower as there were patches of soft powder blue amongst them. I then saw that the blue colour moved, and on closer inspection I discoverd hundreds of tiny blue butterflies feeding off the flowers. I also noticed small fawn and white speckled moths with curved wings and, where small puddles had formed, turquoise and pale blue dragonflies hovered.

I stood enchanted in the silence as I watched the carpet beneath my feet change pattern as first one cloud of fluttering wings and then another moved from flower to flower.

A clanking of metal brought me back to reality and the sight of two workmen fitting a gate to the entrance of the track to seal it off. I ran back shouting to them to let me out. The expression on their faces clearly asked what sort of nut-case would want to walk around such a derelict place. I said nothing about the treasure I had found, but as I walked back to *Bix* I was happy to feel that the flowers and butterflies would remain unmolested in their secret home.

The rest of the year floated peacefully by for us and soon we were starting our seventh year as water gypsies; a national newpaper once described us as such and I must admit I was annoyed when I read the phrase. Alas it was only too accurate because we have heard officially that we are classified now as vagrants. We have no fixed abode, cannot claim Social Benefit if in need, and worst of all, something that I find hard to stomach, we are not even allowed a vote.

But canals, after all, are both the home and the life we chose. We voted for *Bix*.

Index